COVID-19 Pandemic Dynamics

Igor Nesteruk

COVID-19 Pandemic Dynamics

Mathematical Simulations

Springer

Igor Nesteruk
Institute of Hydromechanics
National Academy of Sciences of Ukraine
Kyiv, Ukraine

Electronic Engineering Department
Igor Sikorsky Kyiv Polytechnic Institute
Kyiv, Ukraine

ISBN 978-981-33-6415-8 ISBN 978-981-33-6416-5 (eBook)
https://doi.org/10.1007/978-981-33-6416-5

This Springer imprint is published by the registered company Springer Nature Singapore Pte Ltd.
The registered company address is: 152 Beach Road, #21-01/04 Gateway East, Singapore 189721, Singapore

To my parents, Kateryna and Georgii, who are bravely fighting the pandemic in isolation

Introduction

The COVID-19 pandemic poses a great threat due to millions of infected people, high mortality, and a very negative impact on the economy. Its detailed investigations are still ahead, but the public is already interested in the duration of the pandemic, the expected number of patients, estimations of quarantine measures, the scale of recurrences, etc. The threats of the COVID-19 pandemic require the mobilization of scientists, including mathematicians familiar with methods of infectious disease simulation.

The more complex the mathematical model, the more unknown parameters it contains, the values of which must be determined using a limited number of observations of the disease over time. Even long-term monitoring of the epidemic may not provide reliable estimates of its parameters due to the constant change in quarantine and testing conditions, in algorithms of isolation of infected persons, in pathogen activity, etc.

Any mathematical modeling of the epidemic dynamics will be of particular value if we make an accurate long-term forecast of its duration and number of diseases using statistics data sets obtained immediately after the outbreak. That is why many authors were trying to predict the COVID-19 pandemic dynamics in many countries and regions. We will not dwell on a detailed analysis of these studies and only note that the correct mathematical simulation of the COVID-19 pandemic is very difficult for at least two reasons.

First, data on the number of cases are clearly incomplete immediately after onset, there are quite long hidden periods. The reason is the large number of asymptomatic patients and the lack of skills to detect a new disease. It must be noted that a large discrepancy between the registered and actual number of cases occurred even for the later periods of the COVID-19 pandemic. Adequate modeling is further complicated by the fact that we do not know when the number of reported cases is approaching the actual number. The second reason for the limited accuracy of long-term forecasts is the constant changes in the conditions of the pandemic (changing quarantine measures, social behavior, virulence of the pathogen, etc.). Therefore, a prediction made using statistics for a certain time period is not suitable for other periods of time.

In this book, we will focus on the simplest models that describe the development of infectious diseases over time. In particular, the initial stages of epidemics in different regions, which are characterized by an exponential increase in the number of cases, will be studied. Simple comparisons of the pandemic dynamics in different countries and its trends will be presented. In these parts of the book, I shall use some results of articles and preprints, written together with Gerhard Demelmair, Ihor Kudybyn, Anatolii Nikitin, and Bohdan Shepetyuk, to whom I am also very grateful for collecting and systematizing statistical information on the number of registered cases of the disease.

In this book, we will also use the classical SIR model, with three differential equations for the evolution of the number of susceptible persons—S; infected, spreading the infection—I and removed persons—R, which is the sum of isolated, immunized, and deceased persons. This model contains only four parameters, the values of which can be estimated using a statistical approach developed and successfully applied for investigations of the mysterious children's disease that occurred in the Ukrainian city of Chernivtsi in 1988.

The SIR model is unable to determine the duration of the incubation period and to predict separately the number of deaths caused by coronavirus, but allows us to make adequate predictions of the duration of epidemics in different countries, estimates of their actual beginning (they may precede the time of registration of the first patient), to calculate the time dependences of the total number of patients $V = I + R$ and I, to estimate the probability of meeting an infected person and the effective reproduction numbers. Corresponding results for the first COVID-19 epidemic waves in mainland China, USA, Germany, the UK, the Republic of Korea, Austria, Italy, Spain, France, the Republic of Moldova, Ukraine, the city of Kyiv and for the whole world are already published in my articles and preprints. In the book, these results are systematized and conclusions are drawn based on current information about the course of the pandemic. In Chap. 7 we will compare and discuss the characteristics of the first waves of the COVID-19 pandemic in different countries and WHO regions.

Constant changes in the pandemic conditions (i.e., in the peculiarities of quarantine and its violation, in situations with testing and isolation of patients, in coronavirus activity due to its mutations, etc.) cause changes in the values of parameters of the mathematical models and lead to new pandemic waves. In particular, in October 2020 we observed a sharp increase in the daily number of new cases in many European countries. We will develop simple methods of detecting these changes and propose a simple method of identifying new pandemic waves. The numerical differentiation of smoothed dependences of the accumulated number of cases allows selecting periods with different values of SIR parameters.

To simulate different pandemic waves (periods with more or less constant values of its dynamics parameters), a general SIR model and its exact solution will be proposed. The identification procedures for the parameters of the general SIR model will be described. The characteristics of several pandemic waves in Ukraine and the world will be calculated and corresponding predictions will be presented.

To have good accuracy of predictions, the pandemic dynamics must be updated with the use of new data sets. Because of this, a simple method to assess the final size and duration of epidemic waves was proposed in Chap. 13 and applied for USA, Germany, the UK, the Republic of Korea, Italy, Austria, Spain, France, the Republic of Moldova, Sweden, Ukraine, the city of Kyiv and the whole world.

I would like to express my sincere thanks to Profs. Dirk Langemann (Technische Universitaet Braunschweig) and Juergen Prestin (Universitaet zu Luebeck) for their support in developing the used optimization approach. I would also like to thank Profs. Alberto Redaelli, Giuseppe Passoni, and Gianfranco Fiore (Politecnico di Milano), Sergei Pereverzyev (RICAM, Linz, Austria) for involving me in very interesting biomedical investigations under the EU-funded Horizon 2020 projects: EUMLS (grant PIRSES-GA-2011-295164-EUMLS) and AMMODIT (grant MSCA-RISE 645672).

Finally, my thanks to Academician Viktor Grinchenko, Profs. Volodymyr Tymofeyev, Alexander Galkin, Pavlo Maslianko, my friends Liudmyla Trotsenko, Maia Troian, Nina Basiuk, Anna Voronka, Daria Cusitcaia, Damien Berezenko, Anatolii Podkur, Volodymyr Borysenko, and Oleksii Rodionov for their support and help in collecting and processing data.

Contents

Chapter 1
Data Used for Calculations, Comparisons, and Verifications

Sources of data sets for COVID-19 pandemic dynamics, some examples and principles of their using will be presented in this chapter.

For calculations, we will use the official data about the accumulated numbers of confirmed COVID-19 cases V_j; $j = 1, 2, 3, \ldots, n$ in different countries and worldwide. The number of observations n may vary for different investigations. The main source of this information is WHO daily situation reports, [1]. The corresponding time moments t_j are measured in days. We will use different zero points for the time, specifying them for every calculation (see examples in Tables 1.1 and 1.2). It must be noted that V_j shows the accumulated number of laboratory confirmed cases at the end of corresponding day. It means that the number of infected people could be much higher at corresponding day, since there were and still are rather long delays between the day of testing and day when the results of tests are ready. For example, in Ukraine this delay was between 1 and 10 days (according to the type laboratory and the length of the queue).

We will use and analyze also the daily data for the number of confirmed cases in mainland China, which origins from National Health Commission (NHC) of the People's Republic of China [2] (see Table 1.2). On February 12, 2020, NHC has added 12,289 new cases (not previously included in official counts) as "clinically diagnosed cases". The cases, reported by this official organization before, have the name of "tested confirmed cases" [2]. To avoid confusing, we will denote "tested confirmed cases" as W_j and the "clinically diagnosed cases" as Q_j; j corresponds to the different time moments t_j (see Table 1.2). A part of the official diagram (its version, presented on February 15, 2020) is shown in Fig. 1.1. In some urgent

I. Nesteruk, *COVID-19 Pandemic Dynamics*,
https://doi.org/10.1007/978-981-33-6416-5_1

Table 1.1 Official cumulative numbers of confirmed cases in the Republic of Korea

Day in February 2020	Time moment t_j	Accumulated number of cases in the Republic of Korea V_j, [1]	Day in March 2020	Time moment t_j	Accumulated number of cases in the Republic of Korea V_j, [1]
17	−1	31	1	12	4212
18	0	51	2	13	4812
19	1	104	3	14	5328
20	2	204	4	15	5766
21	3	346	5	16	6284
22	4	602	6	17	6767
23	5	763	7	18	7134
24	6	977	8	19	7382
25	7	1261	9	20	7513
26	8	1766	10	21	7755
27	9	2337	11	22	7869
28	10	3150	12	23	7979
29	11	3736	13	24	8086
–	–	–	14	25	8165

The information from [1]. The corresponding time moments t_j and the accumulated confirmed numbers of cases V_j in South Korea

situations (e.g., after the epidemic outbreak in Italy in February 2020), the information from media [3, 4] was used. The number of cases in Europe was taken from [5]. The accumulated number of cases in Wuhan (China) in December 2019 and January 2020 was calculated with the use of daily data reported in [6] (see Table 1.2).

Some data sets or their parts will be used for calculations, some of them only for verifications of calculations and comparisons. For example, most of the values W_j and $W_j + Q_j$ shown in Table 1.2 were used in two SIR simulations of epidemic dynamics in mainland China, some of them to verify the reliability of these calculations, and the values V_j (also listed in Table 1.2) only for comparisons. All data used for calculations will be given in the relevant tables.

Table 1.2 Data sets used for calculations, comparisons and verifications

Day in December 2019 and January 2020	Time moment t_j	Accumulated number of cases V_j, [6]	Day in January and February 2020	Time moment t_j	"Tested confirmed cases" W_j, [2]	Day in February and March 2020	Time moment t_j	The sum of "tested confirmed cases" and "clinically diagnosed cases" $W_j + Q_j$, [2]
8	−39	1	16	0	45	12	27	58,761
10	−37	2	17	1	62	13	28	63,851
13	−34	3	18	2	121	14	29	66,492
15	−32	5	19	3	198	15	30	68,500
16	−31	6	20	4	291	16	31	70,548
17	−30	7	21	5	440	17	32	72,436
19	−28	9	22	6	571	18	33	74,185
20	−27	13	23	7	830	19	34	74,576
21	−26	17	24	8	1287	20	35	75,465
22	−25	20	25	9	1975	21	36	76,288
23	−24	24	26	10	2744	22	37	76,936
25	−22	28	27	11	4515	23	38	77,150
26	−21	29	28	12	5974	24	39	77,658
27	−20	32	29	13	7711	25	40	78,064
28	−19	34	30	14	9692	26	41	78,497
29	−18	40	31	15	11,791	27	42	78,824
30	−17	45	1	16	14,380	28	43	79,251
31	−16	47	2	17	17,205	29	44	79,824
1	−15	57	3	18	20,440	1	45	80,026
2	−14	64	4	19	24,324	2	46	80,151
3	−13	77	5	20	28,018	–	–	–

(continued)

Table 1.2 (continued)

Day in December 2019 and January 2020	Time moment t_j	Accumulated number of cases V_j, [6]	Day in January and February 2020	Time moment t_j	"Tested confirmed cases" W_j, [2]	Day in February and March 2020	Time moment t_j	The sum of "tested confirmed cases" and "clinically diagnosed cases" $W_j + Q_j$, [2]
4	−12	90	6	21	31,161	–	–	–
5	−11	105	7	22	34,568	–	–	–
6	−10	124	8	23	37,198	–	–	–
7	−9	155	9	24	40,171	–	–	–
8	−8	199	10	25	42,638	–	–	–
–	–	–	11	26	44,653	–	–	–

The information from [2] and [6]. The corresponding time moments t_j and the numbers of V_j, W_j and $W_j + Q_j$

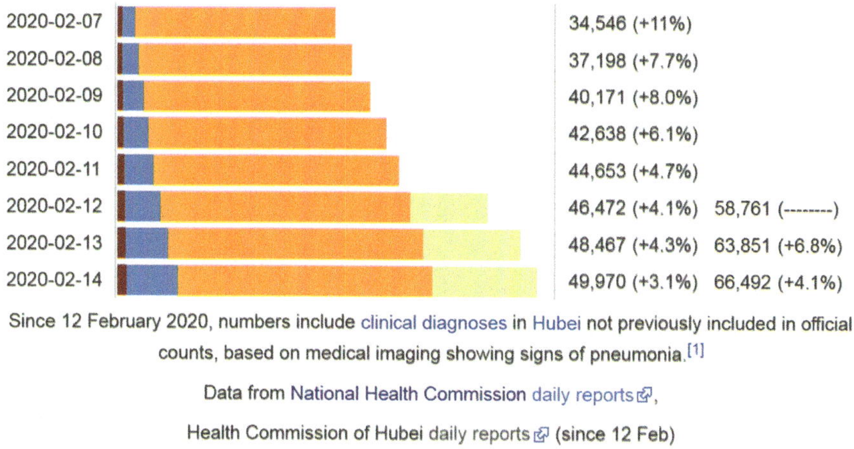

Fig. 1.1 Part of official diagram with the numbers of $W_j + Q_j$ (last column) and W_j (previous column), [2]

Chapter 2
Early Stages of Epidemics and Exponential Growth

The early stages of the COVID-19 epidemic outbreaks and the pandemic dynamics in March and April 2020 will be analyzed. The statistics-based method of parameter estimations for the exponential growth will be presented.

For the initial stage of every epidemic, the exponential growth in the number of cases V over time t is typical:

$$V = e^{\gamma t + \beta} \tag{2.1}$$

where γ and β are constant parameters. This fact was confirmed also for COVID-19 pandemic (see, e.g., [6–32]) and follows from the simple mathematical model, stating that the increase in the number of cases $\mathrm{d}V$ during the time interval $\mathrm{d}t$ is proportional to their number V:

$$\mathrm{d}V = \gamma V \mathrm{d}t \tag{2.2}$$

Formula (2.1) yields the solution to the differential Eq. (2.2) and means that corresponding points V_j (taken in logarithmic scale) versus time must follow the straight lines. This simple model does not take into account the restricted volume of population (according to Eq. (2.1) the number of cases tends to infinity as time increases) and the fact that some infected persons become isolated or dead (i.e., they stop to spread the infection). To simulate the deviation from the exponential growth (2.1) and stabilization of the number of cases at some saturation level, more complicated approaches are necessary. We will use two of them (classical and generalized SIR models) later.

Equation (2.1) yields the linear dependence

$$y \equiv \ln V = \gamma t + \beta \tag{2.3}$$

I. Nesteruk, *COVID-19 Pandemic Dynamics*,
https://doi.org/10.1007/978-981-33-6416-5_2

and the time of cases duplication:

$$\tau_{\mathrm{d}} = \frac{\ln 2}{\gamma} \tag{2.4}$$

To estimate the values of parameters γ and β, we can treat the values $y_j = \log(V_j)$ and corresponding time moments t_j as random variables and use the observations of the accumulated number of cases (e.g., presented in Table 1.1) and the linear regression [33] in order to calculate the coefficients $\widehat{\gamma}$ and $\widehat{\beta}$ of the regression line

$$\widehat{y} = \widehat{\gamma} t + \widehat{\beta} \tag{2.5}$$

using the standard formulas (see, e.g., [33]):

$$\widehat{\gamma} = \frac{n\sum_{j=1}^{n} y_j t_j - \left(\sum_{j=1}^{n} y_j\right)\left(\sum_{j=1}^{n} t_j\right)}{n\sum_{j=1}^{n} t_j^2 - \left(\sum_{j=1}^{n} t_j\right)^2}; \tag{2.6}$$

$$\widehat{\beta} = \frac{\sum_{j=1}^{n} y_j - \widehat{\gamma}\sum_{j=1}^{n} t_j}{n} \tag{2.7}$$

Here n is the number of observations.

Values $\widehat{\gamma}$ and $\widehat{\beta}$ can be treated as statistics-based estimations of parameters γ and β from relationships (2.1) or (2.3). The reliability of this estimation can be checked by calculating the correlation coefficient r:

$$r = \frac{n\sum_{j=1}^{n} y_j t_j - \left(\sum_{j=1}^{n} y_j\right)\left(\sum_{j=1}^{n} t_j\right)}{\sqrt{\left[n\sum_{j=1}^{n} y_j^2 - \left(\sum_{j=1}^{n} y_j\right)^2\right]\left[n\sum_{j=1}^{n} t_j^2 - \left(\sum_{j=1}^{n} t_j\right)^2\right]}} \tag{2.8}$$

If the values of $|r|$ are close to unit, there is a strong linear relationship between variables y and t.

We can use also the F-test for the null hypothesis that says that the proposed linear relationship (2.3) fits the data set. The experimental value of the Fisher function can be calculated with the use of the formula:

$$F = \frac{r^2(n-m)}{(1-r^2)(m-1)} \tag{2.9}$$

where $m = 2$ is the number of parameters in the regression equation, [33]. The corresponding experimental value F has to be compared with the critical value $F_C(k_1, k_2)$ of the Fisher function at a desired significance or confidence level ($k_1 = m - 1$, $k_2 = n - m$; see, e.g., [34]). This approach was successfully used in [35] in order to find the duplication rate for the Chernivtsi children disease. Similar statistics-based approach was applied in [36, 37] to link animal speeds with tail beat amplitude, frequency and body length.

In [38], this method was used to calculate COVID-19 cases duplication time for European region, USA, and the world. The corresponding accumulated numbers of cases used for calculations and comparisons are shown in Table 2.1 and Fig. 2.1 ("circles"). Since in March 2020, the situations in mainland China and the Republic of Korea were stable (the number of new cases was low), and the estimations of global pandemic prospects have been done without taking into account the cases in these two regions. The corresponding numbers V_{Gj} are shown in the last column of Table 2.1 and in Fig. 2.1 (magenta "circles"). The number of cases V_{Ej} and V_{Uj} is also shown in Fig. 2.1 (brown and red "circles," respectively). It can be seen that V_{Ej}, V_{Uj} and V_{Gj} numbers follow straight lines in the logarithmic scale. It means that the epidemic dynamics in these regions was still exponential and far from stabilization (in comparison with Italy—"blue" circles).

To estimate the slopes of these lines, the linear regression for the values $\log(V_{Yj}), Y = E, U, G$ was used (see, (2.3)). The corresponding best fitting dashed lines are shown in Fig. 2.1. To avoid influence of the initial epidemic outbreak stages, calculations have been done only for $V_{Yj} > 100, Y = E, U$; repeating values were used only once. It can be seen that the most rapid was the epidemic in USA (the number of cases duplicated every 2.31 days). The European and global duplications rates were 2.91 and 3.65 days, respectively. Without stabilization by mid-April 2020 (22 days after March 24), we could have in Europe 41 million cases, 38 millions in USA and 211 millions globally. Fortunately, the tendencies changed in April 2020, and these sad predictions became no more relevant.

To illustrate these stabilization tends, we will use some results from [39–41]. The accumulated number of cases confirmed in March and April 2020 is shown in Figs. 2.2, 2.3, 2.4 and 2.5. We use the logarithmic scale to detect the periods of exponential growth which is typical for initial stage of every epidemic. Figure 2.2 demonstrates the pandemic dynamics in the European region, USA and in the world. It can be seen that in April 2020 the European and global dynamics (without cases in mainland China and the Republic of Korea) and the number of cases in USA showed deviation from the straight lines, i.e., the pandemic was stabilizing.

Many countries in Europe demonstrated clear stabilization. For example, we can see it in Fig. 2.3, where the dynamics in Austria and its neighboring countries is shown. Unfortunately, the number of cases in Italy, Spain, Germany, France, and the UK was already higher than in China; in Turkey, Austria, and Switzerland it was higher than in South Korea (see the horizontal lines in Fig. 2.2). In Slovenia, Hungary, and Slovakia, the final number of cases is expected to be rather low. It looks that quarantine measures and isolation of infected persons were very effective

Table 2.1 Official cumulative numbers of confirmed cases in Italy, European region, USA and in the world (without mainland China and South Korea), [1, 38]

Day in February and March, 2020	Time moments in days t_j	Number of cases in Italy V_j	European Region V_{Ej}	USA, V_{Uj}	Global without cases in mainland China and South Korea V_{Gj}
21	−1	9	54	35	1056
22	0	76	121	35	1167
23	1	124	169	35	1306
24	2	229	279	53	1482
25	3	322	379	53	1657
26	4	400	486	59	1898
27	5	650	798	59	2354
28	6	888	1107	62	2859
29	7	1128	1457	62	3433
1	8	1689	2136	62	4562
2	9	2036	2732	64	5753
3	10	2502	3367	108	7341
4	11	3089	4307	129	8993
5	12	3858	5820	148	11,197
6	13	4636	7491	213	14,347
7	14	5883	9453	213	17,593
8	15	7375	12,242	213	21,291
9	16	9172	15,130	472	25,265
10	17	10,149	18,124	696	29,609
11	18	12,462	23,112	987	36,410
12	19	15,113	28,892	1264	43,788
13	20	17660	36,263	1678	53,427
14	21	21,157	45,060	1678	64,304
15	22	24,747	55,623	1678	78,197
16	23	27,980	64,135	3503	89,680
17	24	31,506	74,703	3536	101,612
18	25	35,713	86,994	7087	120,173
19	26	41,035	104,425	10,442	144,121
20	27	47,021	128,434	15,219	175,858
21	28	53,578	151,173	15,219	201,747
22	29	59,138	171,226	31,573	242,368
23	30	63,927	195,261	42,164	281,973
24	31	69,176	220,249	51,914	323,194

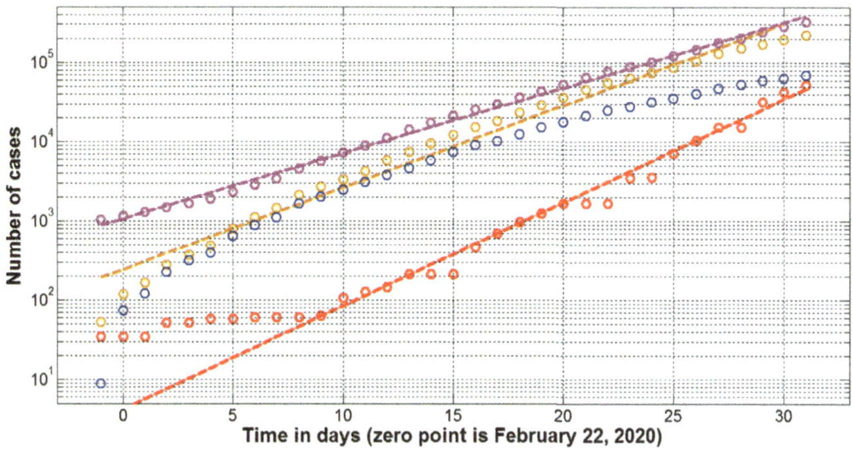

Fig. 2.1 Cases in Italy (blue), European region (brown), USA (red) and in the world (magenta)

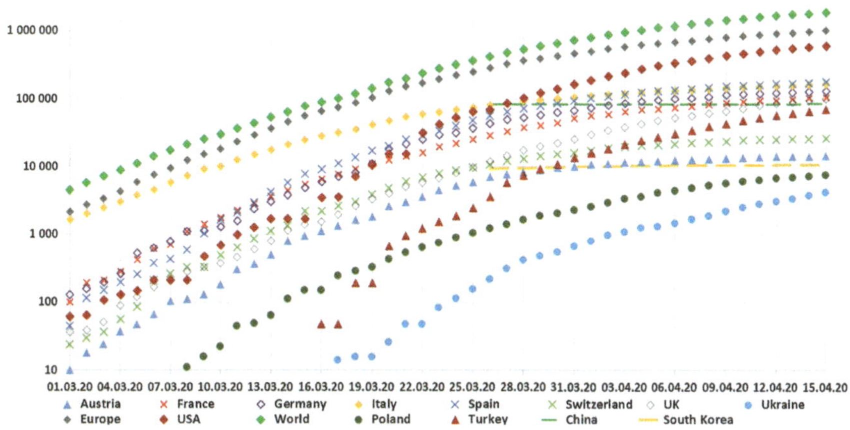

Fig. 2.2 Data sets for the accumulated confirmed number of cases versus time in days in March and April 2020

in these countries. Figure 2.4 shows the pandemic dynamics in Ukraine and its neighboring countries. An evident stabilization is visible only in Poland, Hungary, Moldova, and Slovakia. Some stabilization is visible also in Ukraine. The epidemics in Turkey, Belarus, and RF were still far from stabilization.

Figure 2.5 shows the pandemic dynamics in March and April 2020 for different regions. It can be seen the stabilization in European, Eastern Mediterranean regions, Americas and Africa. But the points for South-East Asia and Western Pacific regions still followed straight lines. Especially threatening was the situation in South-East Asia, where the exponential growth was rather rapid. Unfortunately, the

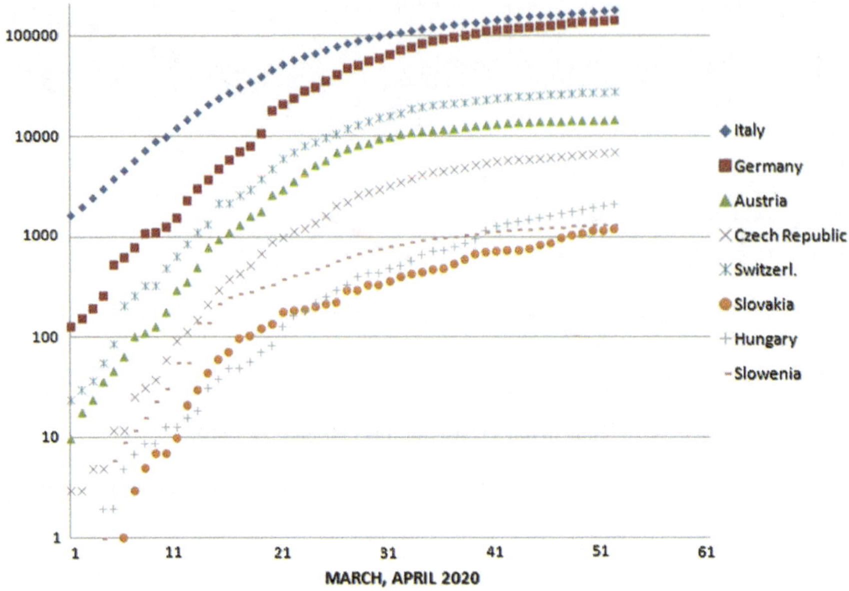

Fig. 2.3 Number of cases in Austria and its neighboring countries versus time in days in March and April 2020

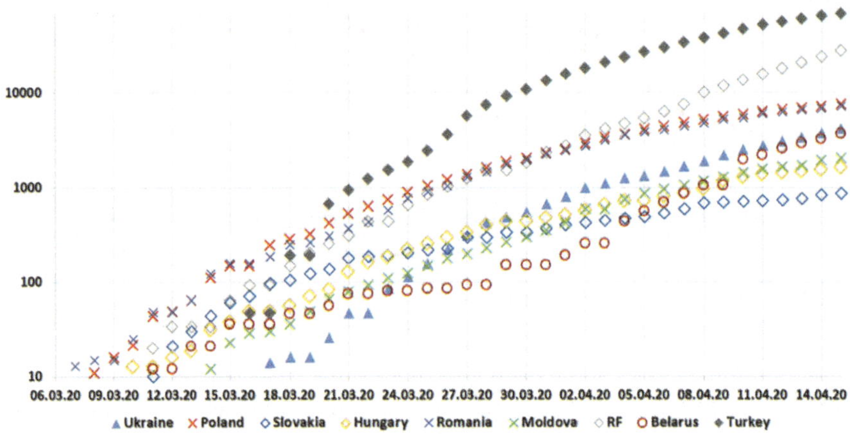

Fig. 2.4 Number of cases in Ukraine and its neighboring countries versus time in days in March and April 2020

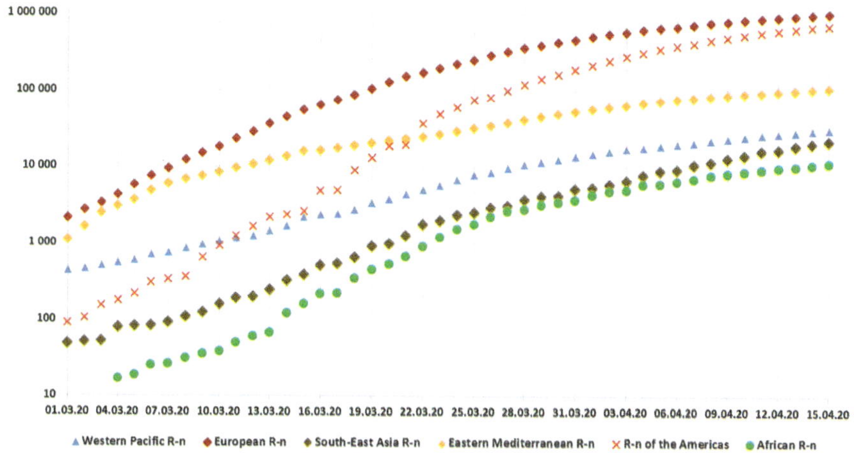

Fig. 2.5 Number of cases in different regions versus time in days in March and April 2020

quarantine weakening, changes in social behavior and maybe in the coronavirus activity caused new rapid increase in the number of cases in August–September 2020. We will discuss these new waves of the pandemic later.

Chapter 3
Comparisons of the Early Stages of the COVID-19 Pandemic in Different Regions

The early stage of the COVID-19 pandemic in Europe is compared with the epidemic dynamics in mainland China. The early stages of the pandemic in Italy, Spain, France, Germany, Switzerland and USA are compared with the dynamics in South Korea. The comparison of the epidemic outbreaks in Ukraine and its neighboring countries is presented.

After the COVID-19 epidemic outbreaks in Italy and other countries, it was very important to compare their dynamics with the cases growth rates in mainland China and the Republic of Korea—regions where the numbers of accumulated cases were high enough to make some statistics-based conclusions. It was also very interesting to compare the initial stages of epidemic outbreaks in different regions and countries [31, 38–51]. In comparison with [31, 44, 46], where rather complicated mathematical simulations were used (in particular, machine learning approaches), we focus here on very simple methods of comparison, which allowed us to get some predictions very quickly. In particular, preprint [42] has been submitted on February 25, 2020 (three days after the epidemic outbreak in Italy).

In [42] and corresponding article [43], the following data for the number of confirmed cumulative cases V_j in Italy were used [3, 4]:

February 22—$V_j = 79$, $t_j = 0$
February 23—$V_j = 152$, $t_j = 1$
February 24—$V_j = 229$, $t_j = 2$
February 25—$V_j = 332$, $t_j = 3$
February 26—$V_j = 400$, $t_j = 4$
February 27—$V_j = 650$, $t_j = 5$
February 28—$V_j = 888$, $t_j = 6$
February 25—$V_j = 332$, $t_j = 3$
February 26—$V_j = 400$, $t_j = 4$
February 27—$V_j = 650$, $t_j = 5$
February 28—$V_j = 888$, $t_j = 6$
February 29—$V_j = 1049$, $t_j = 7$

I. Nesteruk, *COVID-19 Pandemic Dynamics*,
https://doi.org/10.1007/978-981-33-6416-5_3

We have used as well the following information about new cases in EU/EEA and UK, [5]:

for the period February 22–February 29, 1058 new cases, $t_j = 7$
for the period February 22–March 1, 1456 new cases, $t_j = 8$

Thus, the day of February 22, 2020, was taken as a zero point for the epidemic outbreak in Italy. To synchronize the time with the epidemic in mainland China, we can use the linear interpolation between W_j values corresponding moments of time $t_j = 1$ and $t_j = 2$ in Table 1.2. Then the moment of time for China, corresponding to the number of cases 79 (zero point for Italy), is equal 1.2881. Therefore, all time points t_j in the fifth column of Table 1.2 must be decreased by this value. Corresponding figures for the epidemic in China are shown in Fig. 3.1 by "circles". "Stars" represent the number of cases for the epidemic in Italy; "triangles" show new cases in EU/EEA and UK. It can be seen that in February 2020 the epidemic dynamics in Italy, Europe and mainland China were rather similar. Unfortunately, further pandemic development in Europe showed much more rapid increase in the number of cases. Some evidences of these dangers were reported in [50] where more precise SIR simulation was used (see Chap. 6).

A quadratic interpolation for the number of cases confirmed in three days in the Republic of Korea (February 17–19, 2020, see Table 1.2) was used in [41, 49] in order to compare with the initial epidemic dynamics in Italy, Germany, France, Spain, Switzerland and the USA. The numbers of cases confirmed in these countries before March 15, 2020, are shown in Table 3.1. The data sets are taken from [1].

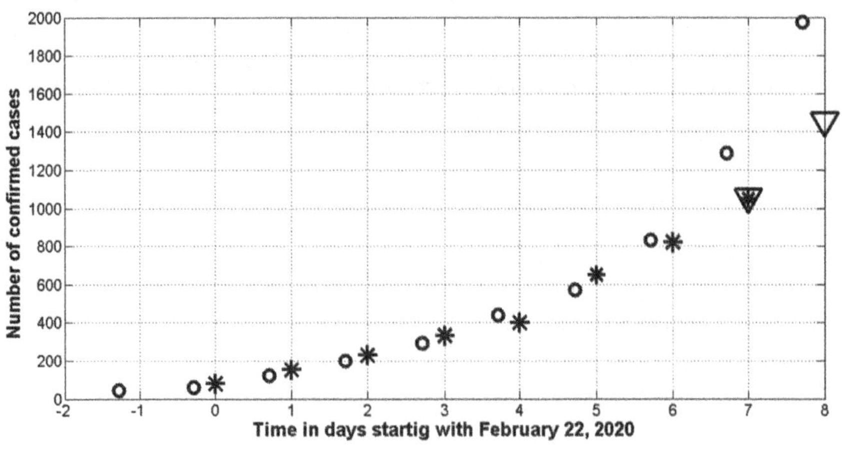

Fig. 3.1 Comparison of the epidemic dynamics in Italy and mainland China. "Circles" represent corresponding W_j values for the epidemic in China (taken from Table 1.2 with the calculated time shift 1.2881); "stars" represent the number of cases in Italy; "triangles" show new cases in EU/EEA and UK confirmed after February 22, 2020

Table 3.1 Number of cases in Italy, Spain, France, Germany, Switzerland and USA used for comparisons and the results of calculations.

Day in February and March, 2020	Italy	Germany	France	Spain	Switzerland	USA	t_{ej}
22	*76*	16	12	2	0	35	0
23	124	16	12	2	0	35	1
24	229	16	12	2	0	53	2
25	322	18	12	2	1	53	3
26	400	21	18	12	1	59	4
27	650	26	38	25	6	59	5
28	888	*57*	57	32	10	62	*6*
29	1128	100	*57*	45	18	62	*7*
1	1689	129	100	*45*	24	62	*8*
2	2036	157	191	114	30	*64*	*9*
3	2502	196	212	151	37	108	10
4	3089	262	282	198	*56*	129	*11*
5	3858	534	420	257	86	148	12
6	4636	639	613	374	209	213	13
7	5883	795	706	430	264	213	14
8	7375	1112	1116	589	332	213	15
9	9172	1139	1402	1024	332	472	16
10	10,149	1296	1774	1639	491	696	17
11	12,462	1567	2269	2140	645	987	18
12	15,113	2369	2860	2965	858	1264	19
13	17,660	3062	3640	4231	1125	1678	20
14	21,151	3795	4469	5753	1359	1678	21
Mortality rate (%), March 14, 2020	6.81	0.21	2.04	2.36	0.81	2.44	–
d_t	0.549	0.154	0.154	−0.179	0.129	0.312	–

The information is about the confirmed accumulated number of cases from [1]. The corresponding time moments t_{ej} are shown in the last column. The values corresponding the beginning of the epidemic outbreak are italicized. The values d_t were calculated with the use of formula (3.2)

Please pay attention that there are some discrepancies in March 22 and 23 figures for Italy shown in the beginning of this chapter.

The mortality rates (shown in Table 3.1) were calculated as of March 14, 2020 (see [1]). It can be seen that the figures vary very significantly from country to country. The mortality rate values also changed greatly over time. Fortunately, world averages tend to decline. For example, as of March 14, 2020, this figure was 3.5%, and in September 2020 it was about 3%, [1]. In this book, we will not focus on the differences in mortalities and their time dependences.

To calculate the number of days from start of the epidemic outbreak which is different for every country (see italicized values in the last column of Table 3.1), it is necessary to compare it with the data set listed in Table 1.1. To increase the accuracy, we let this number to be non-integer. To calculate the corresponding time difference d_t, the parabolic interpolation for the initial number of cases V in South Korea will be used. The first three points in Table 1.1 yield the corresponding equation:

$$V = at^2 + bt + 51, \quad a = 16.5,\ b = 36.5 \tag{3.1}$$

Then by putting into (3.1) the values $V = V_b$ for the number of cases at the starting day of the epidemic outbreak (italicized in Table 3.1), the corresponding values d_t were calculated for every country with the use of formula

$$d_t = \frac{-b + \sqrt{b^2 - 4ac}}{2a}, \quad c = 51 - V_b \tag{3.2}$$

which yields a solution to the quadratic Eq. (3.1). The calculated values of d_t are shown in the last row of Table 3.1. Then the time moments for Italy have to be shifted by 0.549 days, for Germany by 6 + 0.154 days and so on.

The data sets from Tables 1.1 and 3.1 synchronized with the use of Eqs. (3.1) and (3.2) are shown in Fig. 3.2 by "stars" for countries in Europe and by "squares" for USA. The data for South Korea are shown by red line with "circles". Blue "stars" represent the sum of cases for five countries in Europe. It can be seen that in March 2020, Italy, Spain, Germany, Switzerland, and France had no chance for rapid stabilization of the number of cases. The situation in USA was not clear. Unfortunately, in this country very rapid increase of the number of cases occurred after March 15, 2020. Blue "stars" in Fig. 3.2 still followed the straight line, therefore the number of cases in Europe increased exponentially between February 22 and March 14, 2020.

The application of the same simple procedure was illustrated for the epidemic outbreak in Austria. The number of cases in this country was $V_b = 66$ on March 6, 2020. Let us suppose that epidemic in this country started on this day. Corresponding values are as follows: $t_{ej} = 13$ (see Table 3.1) and $d_t = 0.3542$ (according to Eq. (3.2)). If we want to know how the situation looked as of March 15, 2020 (ninth day of the epidemic in Austria), we may open the corresponding WHO report (number 56) and find the number 959 for Austria. This figure has to be compared with the number of cases from Table 1.1, corresponding to the time moment 9 + 0.3542. To avoid any additional calculations, let us take the value 2337 for comparison. Thus, the situation in Austria did not look bad. Further pandemic development was determined by many factors (in particular, the weakening of quarantine and changes in social behavior). As of September 26, 2020 (see WHO weekly update report [52]), the number of cases in Austria (42,940) was much higher than in South Korea (23,611).

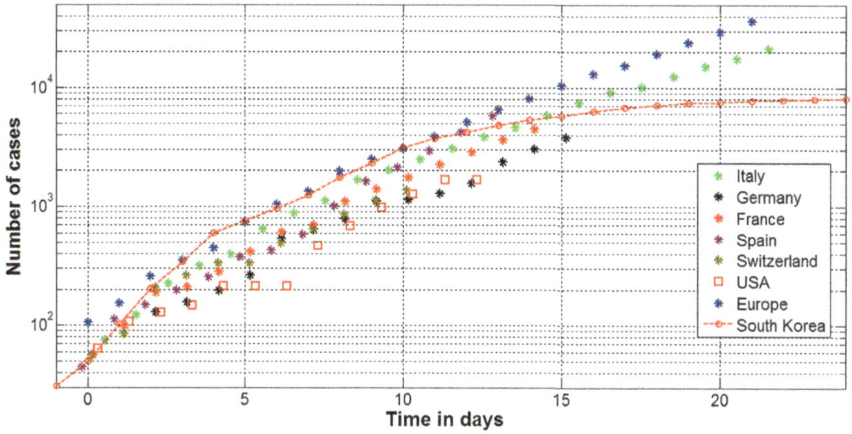

Fig. 3.2 Synchronized data sets for the accumulated confirmed number of cases versus time in days from the beginning of the epidemic outbreaks

We have used the official data sets from [1] to compare the pandemic dynamics in Ukraine and its neighboring countries (see Table 3.2). The day of the first case appearance was taken as the zero day of the epidemic in each country. This simple method allows comparing epidemic dynamics in different countries with different days of the epidemic outbreak. The results are shown in Figs. 3.3 and 3.4 with the use of different scales. "Stars" represent the accumulated number of cases for EU countries; "circles" show the figures for other countries. Figure 3.3 illustrates that the highest number of cases was confirmed in Poland, Romania and Russian Federation (RF). We can compare the number of cases for the fixed day from the epidemic outbreak (e.g., 14th day, when the information from all the countries is available). The most rapid increase occurred in Poland, Slovakia, and the Republic of Moldova.

If we assess the situation in Ukraine only by the number of cases on the 14th day of the epidemic, it did not look so alarming. But the use of the logarithmic scale shows that the number of cases in Ukraine, the Republic of Moldova, Belarus and RF was still growing exponentially (the reference points in Fig. 3.4 follow straight lines). In Ukraine, this growth was the fastest. The further development of the pandemic showed that it was these exponential increases that led to a large number of COVID-19 cases in Ukraine, Moldova, Belarus and RF. As of September 26, 2020 (see WHO weekly update report [52]), the number of cases in Ukraine (198,634) exceeded those in Romania (121,235), Poland (85,980), Belarus (77,289), and Moldova (50,534). As of September 26, numbers of cases in Hungary (24,014) and Slovakia (8600) were the lowest, and in RF (1,151,438) this figure was the highest. The relatively slow epidemic development in Ukraine, Moldova, Belarus, and RF between March 3 and March 21, 2020, can be explained by the fact

Table 3.2 Official cumulative numbers of confirmed cases in Ukraine, Moldova, Poland, Slovakia, Hungary, Romania, Belarus and RF, [1]

Day in February and March 2020	Ukraine	Slovakia	Hungary	Moldova	Romania	Poland	Belarus	RF
21	0	0	0	0	0	0	0	2
22	0	0	0	0	0	0	0	2
23	0	0	0	0	0	0	0	2
24	0	0	0	0	0	0	0	2
25	0	0	0	0	0	0	0	2
26	0	0	0	0	1	0	0	2
27	0	0	0	0	1	0	1	2
28	0	0	0	0	3	0	1	2
29	0	0	0	0	3	0	1	2
1	0	0	0	0	3	0	1	2
2	0	0	0	0	3	0	1	3
3	1	0	0	0	4	1	1	3
4	1	0	2	0	4	1	6	3
5	1	0	2	0	6	1	6	4
6	1	1	5	0	7	5	6	7
7	1	3	7	1	13	6	6	7
8	1	5	9	1	15	11	6	7
9	1	7	9	1	15	16	6	7
10	1	7	13	3	25	22	9	7
11	1	10	13	4	48	44	12	20
12	3	21	16	4	48	49	12	34
13	3	30	19	8	64	64	21	34
14	3	44	32	12	123	111	21	34
15	3	61	39	23	158	150	36	63
16	7	72	50	29	158	150	36	93
17	14	97	50	30	184	246	36	93
18	16	105	58	36	246	287	46	147
19	16	123	73	49	260	325	46	199
20	26	137	85	66	308	425	57	253
21	47	178	131	80	367	536	76	306

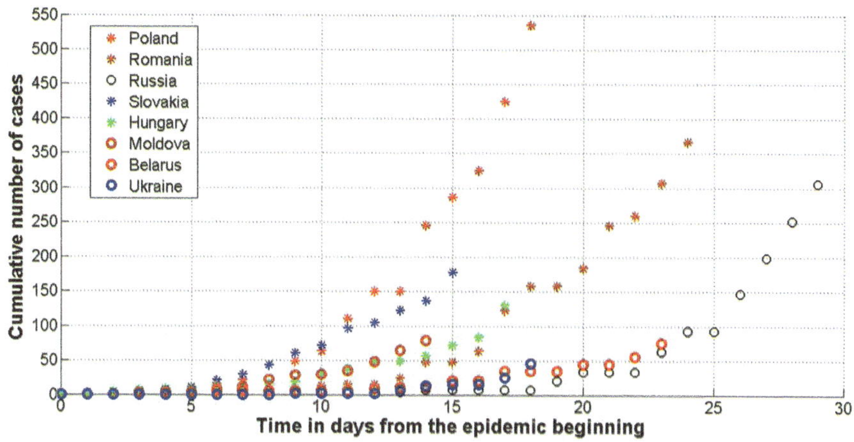

Fig. 3.3 Synchronized data sets for Ukraine and its neighboring countries versus time in days starting with the day when the first COVID-19 case was confirmed (linear scale)

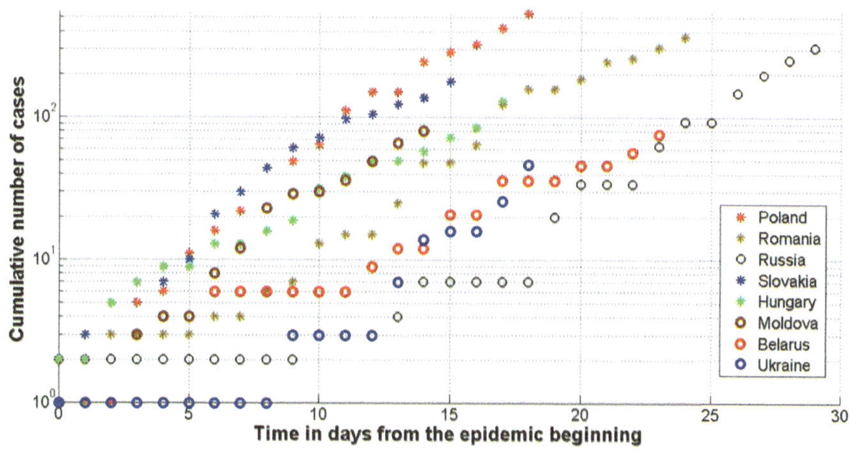

Fig. 3.4 Synchronized data sets for Ukraine and its neighboring countries versus time in days starting with the day when the first COVID-19 case was confirmed (logarithmic scale)

that many infected people were not detected. For example, in Ukraine there were no proper PCR testing (this country started to test the pneumonia patients and the medical staff only after April 10, 2020).

Chapter 4
Classical SIR Model and the Exact Solution of Differential Equations

Differential equations, initial conditions, exact and approximate solutions for the classical SIR model are presented. Simple formulas for probability of meeting an infected person and calculations of the effective and basic reproduction numbers are given.

The SIR model for an infectious disease can be written as follows [53–56]:

$$\frac{dS}{dt} = -\alpha SI \tag{4.1}$$

$$\frac{dI}{dt} = \alpha SI - \rho I \tag{4.2}$$

$$\frac{dR}{dt} = \rho I \tag{4.3}$$

Here t is time; the number of susceptible persons is S (persons who are sensitive to the pathogen and **not protected**); the number of infected is I (persons who are sick and **spread the infection**; please do not confuse with the number of still ill person, so known active cases); the number of removed is R (persons who **do not spread the infection** anymore, this number is the sum of isolated, recovered, dead people and persons who leaved the region); α and ρ are constants measured in $[\text{day}]^{-1}$ if time is measured in days. In some books and papers, the constant $\bar{\alpha} = \alpha N_{pop}$ (N_{pop} is the number of persons in a country or in a region, i.e., the volume of population) is used instead of α. If the value N_{pop} is known, the use of α or $\bar{\alpha}$ is equivalent.

The parameter α is called the infection rate, since according to (4.1) it shows how quick the susceptible persons become infected. Large values of this parameter correspond to severe epidemics with many victims. This parameter accumulates many characteristics. First it shows how strong (virulent) is the pathogen and what is the way of its spreading. For airborne droplets transmission (typical for

I. Nesteruk, *COVID-19 Pandemic Dynamics*,
https://doi.org/10.1007/978-981-33-6416-5_4

COVID-19), the values of α are higher than for AIDS, for example. Parameter α accumulates also the frequency of contacts and the way of contacting. Epidemics used to start at large cities where the average contact rate is much higher than in small villages. So the COVID-19 epidemic outbreak in densely populated regions of Italy is not surprising. In order to decrease the value of α, we have to minimize the number of our contacts and change our contacting habits. For example, we have to avoid the public places and use masks there, minimize or cancel traveling. We have to change our contact habits: to avoid handshakes and kisses. First, all these simple things are very useful to protect yourself. In addition, if the majority of people will do this jointly, we have chance to diminish the value of parameter α and reduce the negative effects of the pandemic.

Usually, the parameter ρ is called the immunization rate. This name is reasonable when there is no isolation and all the sick persons recover. Since the removed persons number R is the sum of isolated, recovered and dead people, Eq. (4.3) demonstrates the increase rate for R. So we will call this parameter removing rate. The inverse value

$$\tau = \frac{1}{\rho} \tag{4.4}$$

is the estimation of the average time of spreading the infection. It is not the average time of the sickness duration. Many people are and will be isolated. They could remain sick and dangerous, but they cannot spread the infection.

So, we are interested in increasing the value of parameter ρ and decreasing the time of spreading the infection $1/\rho$. Public authorities should work on this and organize immediate isolation of suspicious cases. People with problems do not need to go to the hospitals themselves and spread the infection on their way. They have to do their best to minimize spreading the infection. All the people must know how to contact hot lines, and special brigades have to come and isolate suspicious and contact persons.

Since the derivative $\mathrm{d}(S+I+R)/\mathrm{d}t$ is equal to zero (it follows from summarizing Eqs. (4.1)–(4.3)), the sum

$$N = S + I + R \tag{4.5}$$

must be constant for all moments of time and can be treated as the amount of susceptible persons before the moment t_1^* of the epidemic outbreak, since $I = R = 0$ at $t < t_1^*$. It must be noted that the constant N is not the volume of population N_{pop}, but only the initial number of persons who are sensitive and not protected to some specific disease.

Unfortunately, the ratio N/N_{pop} may be very large for the COVID-19 pandemic. We can see this from the situation with the liner Diamond Princess. The total initial number of persons on board was 3711; the final cumulative number of confirmed cases was 741, [52]. It means that the percentage of susceptible can be estimated by

20%, and if the people will be not protected and isolated enough, hundred millions in the world can be infected. The global mortality rate in September 2020 was around 3%, [52]. It means, we have to work hard on protection and isolation to avoid millions of deaths and reduce the number N.

Now it is possible to formulate the recommendations following directly from the presented SIR model:

1. ***Do not panic. Together we can fight the epidemic!***
2. *Minimize the number of contacts and trips, not visit crowded places. Work and study remotely where possible.*
3. *Refrain from shaking hands and kisses during meetings. Use masks in crowded areas. Refrain from having sex with people you do not know (more than half of those infected have no marked symptoms).*
4. *If you (or others) have any suspicious symptoms, do your best to avoid the spread of the infection. Call a dedicated service that should isolate and verify all suspects and contacts.*
5. *Authorities should organize such services and disseminate information as far as possible about symptoms of the disease, where and how to contact them when necessary.*

Unfortunately, too many people do not follow these simple and not very burdensome recommendations, which increases the number of cases (we see a particularly threatening increase in September and October 2020). Moreover, there are a large number of active corona skeptics who organize mass protests against quarantine measures. Mostly, they are young people, often inspired by hostile propaganda. The disease in young people can be mild, often without symptoms. But we must not forget that children and young people can bring the infection to their older relatives and friends, who may die. It is clear that strict quarantine measures are detrimental to the economy, but in a pandemic we cannot live as we did before. It is now the task of governments and society to develop optimal strategies to contain the pandemic without significant economic losses.

The SIR model is the simplest one. It is unable to determine the duration of the incubation period (for this purpose a susceptible-exposed-infectious-recovered (SEIR) model can be used, see, e.g., [7, 57]) and to predict the number of deaths caused by coronavirus. More complicated approaches can be found also in [58–65], but the advantage of SIR model is small number of differential equations and unknown parameters. This fact allows estimating the values of parameters with the use of limited data volumes and making adequate predictions of the duration of epidemics in different countries, estimates of their actual beginning (they may precede the time of registration of the first patient), to calculate the time dependence of the total number of patients $V = I + R$ and I, to estimate the probability of meeting an infected person and the effective reproduction numbers. Corresponding results for mainland China, USA, Germany, the UK, the Republic of Korea, Italy, Austria, Spain, France, the Republic of Moldova, Ukraine, and the city of Kyiv will be presented in next chapters.

To determine the initial conditions for the set of Eqs. (4.1)–(4.3), let us suppose an epidemic started at some moment of time t_1^*, when the first infected person appeared, [35]:

$$I(t_1^*) = 1,\ R(t_1^*) = 0,\ S(t_1^*) = N - 1 \tag{4.6}$$

There are situations when an epidemic starts with several or many victims. For example, when many infected people arrived in the region, the initial value I_1 and $S_1 = N - I_1$ must be used in (4.6).

Very important properties of an epidemic can be derived from the set of differential Eqs. (4.1)–(4.3) without solving; see [53, 54]. In particular, it follows from (4.1) and (4.2) that

$$\frac{\mathrm{d}I}{\mathrm{d}S} = \frac{v}{S} - 1,\quad v = \frac{\rho}{\alpha} \tag{4.7}$$

Integration of (4.7) with the initial conditions (4.6) yields

$$I = v \ln S - S + N - v \ln(N - 1) \tag{4.8}$$

Function I has a maximum at $S = v$ (it follows from (4.7)) and tends to zero at infinity; see [53, 54]. The corresponding number of susceptible persons at infinity $S_\infty > 0$ can be calculated from the following nonlinear equation, [35]:

$$S_\infty = (N - 1)e^{\frac{S_\infty - N}{v}} \tag{4.9}$$

Formula (4.9) follows from (4.8) at $I = 0$ and coincides with the corresponding relationship from [53, 54].

An approximate solution of (4.1)–(4.3) was found by Kermack and McKendrick [53], an exact solution was proposed by Kendall (see [55]). In [35, 66, 67] the set of differential Eqs. (4.1)–(4.3) was solved by introducing the function

$$V(t) = I(t) + R(t), \tag{4.10}$$

corresponding to the number of victims or accumulated number of confirmed cases.

For many epidemics (including the COVID-19 pandemic), we cannot observe dependencies $S(t)$, $I(t)$ and $R(t)$ but observations of the accumulated number of cases V_j corresponding to the moments of time t_j provide information for direct assessments of the dependence $V(t)$. For example, this information is available in Tables 1.1, 1.2, 2.1 and 3.1. We will use also many other data sets in order to estimate the dependence $V(t)$ and values of SIR model parameters.

It follows from (4.2) and (4.3) that

$$\frac{dV}{dt} = \alpha SI \tag{4.11}$$

Then Eqs. (4.5), (4.8) and (4.11) yield

$$\frac{dV}{dt} = \alpha(N - V)[v \ln(N - V) + V - v \ln(N - 1)] \tag{4.12}$$

Finally, after integration of (4.12), we can obtain

$$t = \frac{F_1^*(V, N, v) + \alpha t_1^*}{\alpha} \tag{4.13}$$

$$F_1^* = \int_1^V \frac{dU}{(N - U)[v \ln(N - U) + U - v \ln(N - 1)]} \tag{4.14}$$

Thus, for every set of parameters N, v, α, t_1^* and a fixed value of V the integral (4.14) can be calculated, and the corresponding moment of time can be determined from (4.13). Then I can be calculated from (4.8) and

$$S = N - V, \ R = V - I \tag{4.15}$$

The final number of victims (final accumulated number of cases) can be calculated from

$$V_\infty = N - S_\infty \tag{4.16}$$

To estimate the duration of an epidemic outbreak, we can use the condition

$$I(t_{\text{final}}) = 1 \tag{4.17}$$

which means that at $t > t_{\text{final}}$ less than one person still spreads the infection.

If $N > > V \geq 1$, the obtained exact solution of the set of differential Eqs. (4.1)–(4.3) can be simplified with the use of two different approximations for the function $\ln(N - U) - \ln(N - 1)$. If we assume that $\ln(N - U) - \ln(N - 1) \approx 0$, then $F_1^* = \ln V/N$ and

$$V = e^{\gamma(t - t_1^*)}, \ \gamma = \alpha N \tag{4.18}$$

According to (4.18), epidemics start exponentially and only two parameters ($\gamma = \alpha N$ and t_1^*) describe the process. We have already used Eq. (4.18) in Chap. 2 (Eq. (4.18) coincides with (2.1) if we put $\beta = \gamma t_1^*$).

To follow next epidemic stages, let us use a more exact approximation:

$$\ln(N - U) - \ln(N - 1) \approx (1 - U)/N.$$

Then the integral (4.14) can be expressed as follows:

$$F_1^* = \frac{N}{v + (N - v)N}\left[\ln\frac{NV + v(1 - V)}{N} - \ln\frac{N - V}{N - 1}\right]$$

and the solution has the form

$$V = \frac{EN - v}{E + N - v} \tag{4.19}$$

$$E = \frac{N}{N - 1}\exp\frac{\alpha\left(t - t_1^*\right)[v + (N - v)N]}{N}$$

The second approximation (4.19) yields limited value of victims, since $V = I + R$ tends to N at infinity, but the differences between the exact solution (4.13)–(4.14) and approximate solutions (4.18) and (4.19) can be very large (see [35]). Further, we will use only the exact solution (4.13)–(4.14) (similar to papers [38, 50, 66–78]).

The constant parameters of the model N, v, α, t_1^* and SIR curves can be calculated with the use of limited number of observations V_j; $j = 1, 2, 3, \ldots, n$ during some limited period of time T_c (a corresponding algorithm will be presented in the next chapter). The obtained SIR curves allow predicting the epidemic behavior (after T_c) and restoring the epidemic history (before T_c). When applying the above formulas, it should be remembered that the model assumes a closed population with stable conditions of the epidemic (because all four parameters N, v, α, t_1^* are assumed to be constant from the time moment t_1^* to the time moment t_{final}). If new infected people arrive in the region (so known imported cases) and/or there are changes in quarantine conditions, testing algorithms or social behavior, the accuracy of both predictions and analysis of the past may deteriorate. Some modified approach will be developed later in order to take into account changes in epidemic characteristics.

Actually, the presented approach is suitable only for the first waves of an epidemic, since the initial condition (4.6) assumes that the epidemic began with one person. In August and September 2020, we saw a sharp increase in the daily number of new COVID-19 cases in many countries, which may indicate a change in the pandemic characteristics. These new waves cannot be adequately described by the above formulas, so to assess the accuracy of the calculations of the first epidemic waves in different countries, we can offer the following formula:

$$\varepsilon = \frac{\left|V_f - V_\infty\right|}{V_f} \tag{4.20}$$

where V_f is the real accumulated number of cases registered at day corresponding to t_{final}.

Indeed, the presented theory only states that for the invariant characteristics of the epidemic (N, v, α, t_1^*) during the time after T_c to t_{final}, the real number of cases at time moment t_{final} should be equal to the calculated value of V_∞. Therefore, we have the right to compare V_f and V_∞. Equation (4.20) is a characteristic not only of the accuracy of the theory, but also an assessment of the fact that in the period after T_c to t_{final} the characteristics of the epidemic changed insignificantly (e.g., the quarantine conditions or social behavior changed little). Chapter 6 will provide examples of the application of the SIR model and calculations of the relative error ε with the use of formula (4.20) for many countries and regions and will discuss the reasons for small and large discrepancies between forecasts and actual observations.

As long as there are infected people in the population, there is a chance to meet one of them. The corresponding probability can be estimated with the use of simple formula:

$$p(t) = \frac{I(t)}{N_{\text{pop}}} \tag{4.21}$$

where N_{pop} is the volume of population. With the use of corresponding SIR curves, it is possible to estimate $p(t)$ for every moment of time. Formula (4.21) can be useful for deciding which countries' citizens are welcome guests after the resumption of international passenger traffic. If $p_A(t) > p_B(t)$ for countries A and B, citizens of A are not welcome guests in country B at the moment of time t.

In April–July 2020, the quarantine weakening and its strengthening in September–October 2020 have become urgent in many countries. The SIR $I(t)$ curves can be used to assess their feasibility. For example, number of infected persons in Kyiv (some of them may feel completely healthy or have mild symptoms) was estimated by 100 on May 11, 2020 [73, 74]. It means that the probability to meet such person was rather low ($p \approx 3,4 \times 10^{-5}$). But if you have M contacts, the probability of meeting at least one infected person increases according to the formula:

$$P = 1 - (1 - p)^M \approx pM \tag{4.22}$$

Therefore, a person on duty in the subway or a bus driver is quite likely to meet an infected person during, for example, 15 days of work. Meeting an infected person does not mean getting infected, so masked mode and distance in transport and other public places must remain mandatory. Workers in transport, trade, pharmacies, police (all of whom are forced to have many contacts) must be provided with enhanced protection. People at risk should continue to refrain from traveling, visiting indoors and minimizing visits to medical facilities. Where it is possible, distant work and study should be maintained. For ordinary citizens working in small groups, the risk of meeting an infected person depends on the transport situation. If cars and metro stations, land transport will be regularly decontaminated, then the values of M and P for passengers will be small. Everyone can assess their own level of risk (allowable probability) using formula (4.22).

According to (4.21) the maximum probabilities of meeting an infected person, $p(t)$ corresponds to the maximum I_{max} of the function $I(t)$ which reached at $S = v$ (see (4.7)). Then Eq. (4.8) yields

$$I_{max} \equiv I(t_{max}) = v \ln \frac{v}{N-1} + N - v \tag{4.23}$$

Function $I(t)$ increases at the time period $t_1^* < t < t_{max}$ and symmetrically decreases at $t_{max} < t < t_{final}$. Duration of the first wave of an epidemic $T_d = t_{final} - t_1^*$ can be calculated with the use of simple formula:

$$T_d = 2\left(t_{max} - t_1^*\right) \tag{4.24}$$

Since $S(t_{max}) = v$(see (4.7)), integration of differential Eq. (4.1) taking into account (4.6), (4.8) and (4.24) yields

$$T_d = -\frac{2}{\alpha} \int_{N-1}^{v} \frac{dU}{U[v \ln U - U + N - v \ln(N-1)]} \tag{4.25}$$

SIR model also allows us to determine the point in time t_s when the daily increase in new cases will begin to decline. The daily increase in the number of cases can be estimated as $\mathrm{d}V/\mathrm{d}t$ and calculated with the use of (4.11) and (4.12). Then the maximum in daily increase corresponds to zero value of the second derivative $\mathrm{d}^2V/\mathrm{d}t^2$. Differentiation of (4.11) yields

$$\frac{\mathrm{d}^2 V}{\mathrm{d}t^2} = \alpha\left(S\frac{\mathrm{d}I}{\mathrm{d}t} + I\frac{\mathrm{d}S}{\mathrm{d}t}\right) \tag{4.26}$$

With the use of (4.1) and (4.2), Eq. (4.26) can be rewritten as follows:

$$\frac{d^2V}{dt^2} = \alpha^2 SI(S - I - v) \tag{4.27}$$

According to (4.27) the maximum of daily cases increase corresponds to the value

$$S_s = I_s + v \tag{4.28}$$

where corresponding value I_s may be calculated from (4.8):

$$I_s = v \ln S_s - S_s + N - v \ln(N - 1) \tag{4.29}$$

Because function S decreases monotonically (see (4.1)), and the maximum of I corresponds to $S = v$, Eq. (4.29) shows that the daily number of new cases starts to decline before the maximum number of the infected (and spreading) persons is achieved. This conclusion is very important, because watching a new epidemic, we record the daily number of new cases or the accumulated number of cases. If the daily amount begins to decline, it does not mean that the number of carriers of the infection is also declining and quarantine can be relaxed. To estimate the corresponding time difference $t_{\max} - t_s$, we continue the analysis of the system of Eqs. (4.1)–(4.3).

Equations (4.28) and (4.29) yield the following nonlinear equation for S_s:

$$S_s = \frac{1}{2}\left(v + v \ln \frac{S_s}{N-1} + N\right) \tag{4.30}$$

After solving (4.30), the point in time t_s can be found by integration (4.1)

$$t_s = t_1^* - \frac{1}{\alpha} \int\limits_{N-1}^{S_s} \frac{dU}{U[v \ln U - U + N - v \ln(N - 1)]} \tag{4.31}$$

Equations (4.24), (4.25), (4.30), (4.31) allow estimating the time difference $t_{\max} - t_s$, but values of the SIR parameters have to be calculated first.

Important epidemic characteristics—the basic and effective reproduction numbers—can be also estimated with the use of SIR model. The effective reproduction number $R_t(t)$ shows the average number of people infected by one person [79]. In terms of SIR model, it can be estimated as follows:

$$R_t(t) = \frac{I(t + 0.5\tau) - I(t - 0.5\tau)}{I} = \frac{1}{I} \int\limits_{t-0.5\tau}^{t+0.5\tau} \frac{dI(t^*)}{dt^*} dt^* \approx \frac{\tau}{I} \frac{dI(t)}{dt} \tag{4.32}$$

where τ is the average time of spreading the infection and can be estimated with the use of (4.4).

The basic reproduction number R_0 for the initial periods of an epidemic outbreak can be estimated with the use of equation, [79]:

$$R_0 = \exp\left(\frac{\tau \ln 2}{\tau_{\rm d}}\right) \tag{4.33}$$

where $\tau_{\rm d}$ is the doubling time (the period during which the number of cases duplicates; the exponential growth is assumed).

Chapter 5
Statistics-Based Procedure of Parameter Identification for the Classical SIR Model

The procedure of the SIR parameter identification and examples of calculations are presented.

In the case of a new epidemic, the values of four independent parameters N, v, α, t_1^* of SIR model are unknown and must be identified with the use of limited data sets. A statistics-based approach was developed in [35] and used in [38, 50, 66–78] to estimate the values of these unknown parameters for different countries and regions. The essence of the method is to use observations of the number of victims V_j corresponding to the moments of time t_j to calculate $F_{1j} = F_1^*(V_j, N, v)$ for every fixed values N and v with the use of (4.14) and then to check how the registered points fit the straight line (4.13).

Equation (4.13) can be rewritten as follows:

$$y \equiv F_1^*(V, N, v) = \alpha t - \alpha t_1^* \tag{5.1}$$

Equation (5.1) coincides with (2.3) if we assume

$$\alpha = \gamma, \ \beta = -\alpha t_1^* \tag{5.2}$$

As in Chap. 2, we can estimate the values of parameters γ and β, by treating the values $y_j \equiv F_1^*(V_j, N, v)$ and corresponding time moments t_j as random variables and use the observations of the accumulated number of cases (e.g., presented in Table 5.1) and the linear regression [33] in order to calculate the coefficients $\widehat{\gamma}$ and $\widehat{\beta}$ of the regression line

$$\widehat{y} = \widehat{\gamma} t + \widehat{\beta} \tag{5.3}$$

using the standard formulas (2.6) and (2.7). Values $\widehat{\gamma}$ and $\widehat{\beta}$ can be treated as statistics-based estimations of parameters γ and β from relationships (2.3). Then the estimations for parameters α, t_1^* can be easily recalculated using (5.2).

I. Nesteruk, *COVID-19 Pandemic Dynamics*,
https://doi.org/10.1007/978-981-33-6416-5_5

Table 5.1 Accumulated number of COVID-19 cases in Ukraine [1] and its capital Kyiv [80] used for calculations and verifications of the results

Day in April and May 2020	Moments of time in days t_j	Accumulated number cases in Kyiv	Accumulated number cases in Ukraine V_j
25	53	1188	8617
26	54	1220	9009
27	55	1298	9410
28	56	1341	9866
29	57	1401	10,406
30	58	1453	10,861
1	59	1494	11,411
2	60	1511	11,913
3	61	1535	12,331
4	62	1583	12,697
5	63	1651	13,184
6	64	1675	13,691
7	65	1721	14,195
8	66	1771	14,710
9	67	1805	15,232
10	68	1871	15,648
11	69	1930	16,023
12	70	1960	16,425
13	71	2012	16,847
14	72	2068	17,330
15	73	2221	17,858
16	74	2280	18,291
17	75	2321	18,616
18	76	2359	18,876
19	77	2410	19,230
20	78	2475	19,706
21	79	2522	20,148
22	80	2569	20,580
23	81	2623	20,986
24	82	2652	21,245
25	83	2710	21,584
26	84	2739	21,905
27	85	2793	22,382
28	86	2849	22,811
29	87	2904	23,294
30	88	2985	23,672
31	89	3043	24,002

The reliability of the method can be checked by calculating the correlation coefficient r with the use of (2.8) and checking how close its value is to unity. We can use also the F-test for the null hypothesis that says that the proposed linear relationship (5.1) fits the data set. The experimental value of the Fisher function F can be calculated using (2.9). The corresponding experimental value F has to be compared with the critical value $F_C(k_1, k_2)$ of the Fisher function at a desired significance or confidence level ($k_1 = m - 1$; $k_2 = n - m$; n is the number of observations; $m = 2$ is the number of parameters in the regression equation, see, e.g., [34]). When the values n and m are fixed, the maximum of the Fisher function coincides with the maximum of the correlation coefficient. Therefore, to find the optimal values of parameters N and v, we have to find the maximum of the correlation coefficient for the linear dependence (5.1) or (5.3). To compare the reliability of different predictions (with different values of n), it is useful to use the ratio $F/F_C(1, n-2)$ at fixed significance level. The most reliable estimation of the SIR model parameters corresponds to the highest $F/F_C(1, n-2)$ value. This ratio was used in [37] to find optimal links between the water animal speeds and their body length. In further calculations, we will use the significance level 0.001; corresponding values of $F_C(1, n-2)$ can be taken from [34].

The calculations of $F_{1j} = F_1^*(V_j, N, v)$ using (4.14), linear regression coefficients (5.3), correlation coefficient r, and $F/F_C(1, n-2)$ ratio were performed with the use of original MATLAB code, which allows us also to isolate the values of parameters N and v corresponding to the maximum of r. The exact solution (4.13)–(4.14) allows avoiding numerical solutions of differential Eqs. (4.1)–(4.3) and significantly reduces the time spent on calculations. For large values of V_j (e.g., for the USA and the world), calculations of integral (4.14) need more computer time, but even in these cases the calculation of one set of optimal parameters could be performed on a regular laptop for one day.

Examples of calculations with the use of the accumulated number of cases, confirmed in Ukraine (see Table 5.1), are presented in Table 5.2. It can be seen that all the predictions yield very high values of the correlation coefficient (close to unit) and high values of the $F/F_C(1, n-2)$ ratio. The results of calculations depend on the number of observations n (compare Predictions 4 and 5) and the data set (at fixed values of n, compare Predictions 5 and 6). This fact reflects both the accuracy of the approach and the changes in the epidemic behavior (for example on May 10, 2020, the national lockdown was canceled and the quarantine restrictions changed permanently).

In SIR model, all the parameters are supposed to be constant. If the quarantine measures and speed of isolation change or new infected persons are coming in the country, the accuracy of the prediction reduces. It is very important to detect the periods of time with more or less constant values of parameters, i.e., to select different waves of the pandemic and to find optimal values of the parameters separately for every wave. These problems will be discussed later.

Without separation of the waves, we will obtain different values of the important epidemic characteristics. For example, the real time of the epidemic beginning t_1^*

Table 5.2 Calculated optimal values of SIR model parameters for the COVID-19 epidemic in Ukraine

Number of prediction/ optimal values of parameters	Prediction 4 $n = 14$ Last point taken for calculations is May 21, 2020	Prediction 5 $n = 21$ Last point taken for calculations is May 21, 2020	Prediction 6 $n = 21$ Last point taken for calculations is May 31, 2020
N	35,492.6848	38,056.3712	50,436.16
v	15,796.5412546773	20,734.7730884932	25,477.9529196339
α	3.426845968144e-06	4.3395273241e-06	1.84021927083728e-06
t_1^*	−69.8107630794681	−53.1438951181010	−98.4845510676046
ρ	0.0541323137092154	0.0899791143765501	0.0468850199441952
$1/\rho$	18.4732543554620	11.1136901816475	21.3287741199694
r	0.999081329709670	0.999638595705623	0.999291586201258
F, Eq. (2.9)	6522.18026093915	26,272.1014852200	13,395.9930093636
$F/F_C(1, n-2)$	350.654852738664	1728.42772929079	881.315329563392
S_∞, Eq. (4.9)	5223	9689	10,533
V_∞, Eq. (4.16)	30,270	28,368	39,903
t_{final}, Eq. (4.17)	358.1	280.8	495.8

varies by 45 days for different predictions (see Table 5.2); the final sizes V_∞ are much smaller than the number of cases in Ukraine in September 2020 (this figure exceeded 200 thousands, [1]). Estimations of the average spreading time $\tau = 1/\rho$ and final days t_{final} (when no new cases are expected) differ by 40%. All the predictions yield rather high values of S_∞. It means that the epidemic in Ukraine was expected to stop with many susceptible persons who could probably catch the infection. Thus, a new wave can start. Unfortunately, in September 2020, these waves were clearly visible in Ukraine and many other countries.

We must understand also that different parameter identification procedures can yield different optimal values of the SIR models parameters (see, e.g., [56]). The best verification methods are to compare calculated SIR curves for $V = I + R$ with V_j values, obtained after the calculations have been completed or/and to use estimation (4.20) (if it is possible). The calculations of this and other SIR curves, the very important predictions of the expected final number of cases and the final day of the epidemic (see last two rows in Table 5.2) were performed with the use of another original MATLAB code. Knowing the optimal values of the parameters N, v, α, t_1^*, anyone familiar with the differential equations can obtain SIR curves and predictions with the use of formulas available in the previous chapter. The results of calculations for different countries and regions will be presented in the next chapter.

Chapter 6
SIR Simulations for the First Waves of the COVID-19 Pandemic in Different Countries and Regions

In this chapter, we will summarize the results of papers [38, 50, 66–78] and present applications of the approach proposed in two previous chapters to simulate the first waves of the COVID-19 pandemic in mainland China, USA, Germany, the UK, the Republic of Korea, Italy, Austria, Spain, France, the Republic of Moldova, Ukraine, the city of Kyiv, and the whole world.

6.1 Mainland China

The author drew attention to the COVID-19 epidemic dynamics in early February 2020. Since the original method of determining the parameters of the SIR model was published in 2018 [35], its application did not take much time, and on February 10, 2020, the first preprint [81] was posted and the manuscript was submitted to the Journal Innovative Biosystems and Bioengineering. It turned out that the technique developed and applied to analyze the dynamics of the mysterious disease of children in the Ukrainian city of Chernivtsi can be useful for mathematical simulations of the COVID-19 pandemic.

Since the real number of cases confirmed in mainland China before February 12, 2020, is still unknown, the values of W_j ("tested confirmed cases" [2], listed in Table 1.2), were used for calculations. Thus, it was supposed that $V_j = W_j$ and the calculations with the use of data for the time period from January 16 to February 9, 2020, and the method presented in Chaps. 4 and 5 have been done. The optimal values of parameters are shown in Table 6.1 (prediction 1).

The corresponding correlation coefficient $r = 0.997966$ was quite high, and the forecast was very optimistic: The final number of cases V_{∞} was predicted to be around 45 thousands, and the new cases will stop to appear after February 25, 2020 ($t_{\text{final}} = 40$, see Table 6.1). The corresponding numbers of infected I, susceptible S and removed R persons versus time (starting from January 16, 2020) are calculated and shown in Fig. 6.1. The blue line represents the number of victims

I. Nesteruk, *COVID-19 Pandemic Dynamics*,
https://doi.org/10.1007/978-981-33-6416-5_6

Table 6.1 Calculated optimal values of SIR model parameters for the COVID-19 epidemic in mainland China

Number of prediction/ optimal values of parameters	Prediction 1 $n = 25$ Period taken for calculations T_c: January 16–February 9, 2020	Prediction 2 $n = 17$ Period taken for calculations T_c: February 13–29, 2020
N	90,611	138,317.9776
v	65,546.5	91,771.3993770612
α	1.477985357571669e-05	4.27635873268949e-06
t_1^*	−7.720998173432072	−32.2443982214093
ρ	0.9688	0.3924
$1/\rho$	1.0322	2.5481
r	0.997966487046645	0.994531292187372
S_∞, Eq. (4.9)	45,579	57,061
V_∞, Eq. (4.16)	45,032	81,257
t_{final}, Eq. (4.17)	40	119
Final day of the first wave	February 25, 2020	May 14, 2020
V_f, taken from [1]	78,191	84,469
ε, Eq. (4.20)	42.4%	3.8%

$V = I + R$ and is in good agreement with "tested confirmed cases" W_j, reported by National Health Commission (NHC) of the People's Republic of China [2].

One can imagine the excitement with which the author expected new reports from NHC in order to verify the results of calculations. The waiting process even became the subject of a separate preprint [68]. Unfortunately, many cases have not be included in the official counts and have appeared in the official table from [2] only on February 12 as "clinically diagnosed cases" Q_j (see Table 1.2 and Fig. 1.1). Since NHC has proposed two different ways of registration of the same disease [2], V_j must be the sum of W_j and Q_j, i.e., $V_j = W_j + Q_j$ (provided that no new methods of registering the same disease would appear). Values W_j after February 9 are shown in Fig. 6.2 by "stars." "Crosses" represent the sum $W_j + Q_j$.

Since the optimal curve was obtained only with the use of W_j and the difference between W_j and V_j is very big (e.g., it was 12,289 persons on February 12, 2020), the prediction 1 shown in Table 6.1, Figs. 6.1 and 6.2 and reported in [66–68, 81] became no longer relevant. To have better estimations, it is necessary to have exact Q_j data for the period before February 12.

Since more than fourteen thousands new cases (not previously included in NHC official counts) have been added on February 12, 2020, a new prediction has been carried out in [69] with the use of official number of cases shown by NHC after February 12. The calculations have been performed with the use of $W_j + Q_j$ values (see last column in Table 1.2) for the period of time February 13–February 29 (the number of observations $n = 17$) and the assumption that $V_j = W_j + Q_j$. The optimal

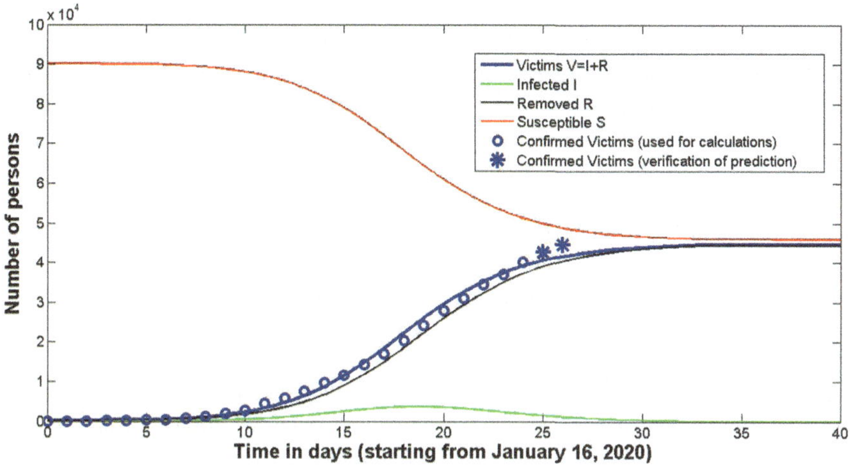

Fig. 6.1 Mainland China, prediction 1. Results of calculations and verification. Numbers of infected I (green line), susceptible S (red line) and removed R (black line) persons versus time in days (starting from January 16, 2020); the blue line represents the number of victims $V = I + R$; blue markers show the "tested confirmed cases" W_j ([2], Table 1.2); the "circles" correspond to the points used for calculations (it was supposed that $V_j = W_j$); "stars"—to the points used only for verification

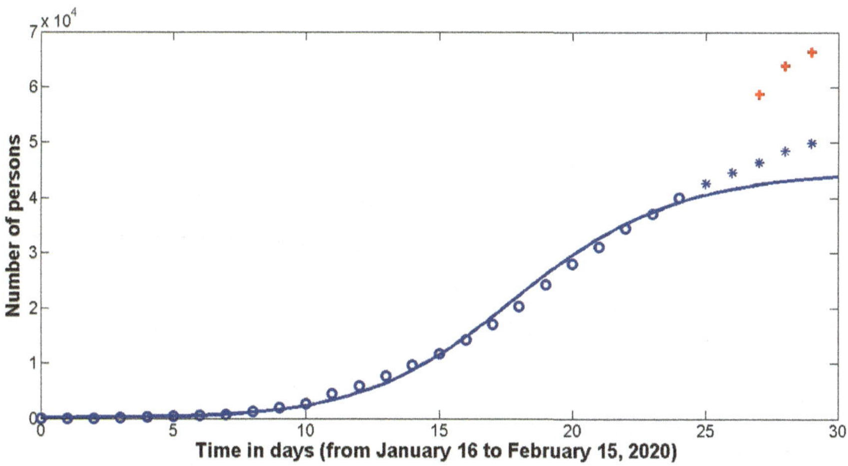

Fig. 6.2 Mainland China, prediction 1 (line) and numbers of cases from NHC [2] (markers). Line represents $V = I + R$ dependence, calculated with the use of optimal parameters shown in Table 6.1 (prediction 1); "circles" show the "tested confirmed cases" W_j for the period from January 16 to February 9, 2020 ([2], Table 1.2) used for calculations; "stars" correspond to the "tested confirmed cases" W_j for the period February 10–14, 2020 [2]; "crosses" represent the sum $W_j + Q_j$ from [2]

values of parameters are presented in Table 6.1 (prediction 2). It can be seen that the value of the correlation coefficient is very high $r = 0.99453$. The value of the Fisher function is 1360.2 (according to (2.9)) that far exceeds the critical value $F_C(1, 15) = 16.6$ even for the significance level 0.001 (see, e.g., [34]).

The values of the final number of susceptible persons $S_\infty \approx 57,061$ and the final number of victims $V_\infty \approx 81,257$ were calculated with the use of (4.9) and (4.16). The corresponding numbers of infected I, susceptible S and removed R persons versus time (starting from January 16, 2020) are presented in Fig. 6.3. The blue line represents the number of victims $V = I + R$ and is in good agreement with $W_j + Q_j$ values taken for calculations (blue "circles"). The difference between the blue line and W_j values (shown by blue "triangles") can be treated as corresponding Q_j values which are not available for statistical analysis.

To estimate the epidemic duration in mainland China (final time moment t_{final}), the condition (4.17) was used. The calculations gave us the value $t_{\text{final}} \approx 119$. According to this prediction, the end of epidemic in China was expected to be around May 14, 2020, provided the existing quarantine measures and rates of patient isolation will continue and the warm weather will not weaken the coronavirus.

To demonstrate the accuracy of prediction 2, two new values of $W_j + Q_j$ corresponding the first two days of March are shown in Fig. 6.3 by blue "stars." As noted in the previous chapters, the applied approach can describe only the first wave of the epidemic and does not take into account imported cases, so to assess its

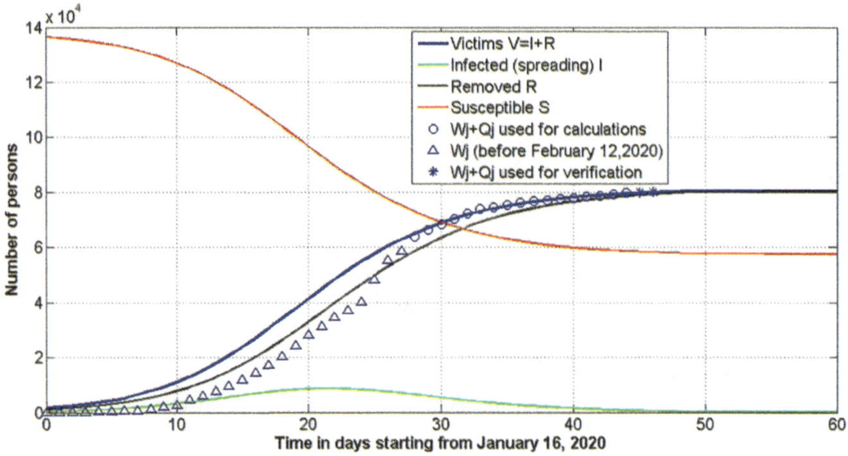

Fig. 6.3 Mainland China, prediction 2. Results of calculations and verification. Numbers of infected I (green line), susceptible S (red line) and removed R (black line) persons versus time in days (starting from January 16, 2020); the blue line represents the number of victims $V = I + R$; blue markers show cumulative number of confirmed cases, reported by NHC [2]; the "circles" correspond to the points used for calculations (it was supposed that $V_j = W_j + Q_j$); "stars"—to the points used only for verification; "triangles" represent the W_j data set, which has not been used for calculations (see Table 1.2)

accuracy; we can use information on the accumulated number of cases V_f recorded in the day, which corresponds to t_{final}. The corresponding values from [1] are shown in Table 6.1. Then, the accuracy ε can be estimated with the use of formula (4.20). The values presented in Table 6.1 demonstrate that the accuracy of the first prediction is rather bad (this is not surprising, as incomplete information on the number of cases was used for the calculations). But the second prediction came true with great accuracy ($\varepsilon = 3.8\%$). As of September 26, 2020, the number of COVID-19 cases in China reached the number 90,966 [52]. It means that for almost 7 months of observation of the epidemic (the second prediction was posted on March 4, 2020), the revealed exceeding the forecast level was only 11%.

Figure 6.4 illustrates the early stage of epidemic according to prediction 2. The numbers of victims summarized with the use of data available in [6] are shown by blue "squares." The prediction 2 demonstrates that possible epidemic beginning happened 25 days earlier (compare t_1^* values for predictions 1 and 2 in Table 6.1). The $V = I + R$ line shown in Fig. 6.4 supports the well-known fact that the first cases of COVID-19 were laboratory confirmed in December 2019 and correlates with the number of cases reported in [6]. Nevertheless, the further investigations demonstrate that the COVID-19 epidemic in China probably stated much earlier. Later, we present the results of calculations with the use of global data set, which will estimate the real pandemic beginning in August 2019.

The values of the parameter $\rho = \nu\alpha = 0.3924$ and the inverse value $1/\rho = 2.5481$ are shown in Table 6.1 (prediction 2). Thus, the average time of spreading the infection can be estimated as approximately 2.5 days and is much higher in comparison with the prediction 1 based on incomplete data.

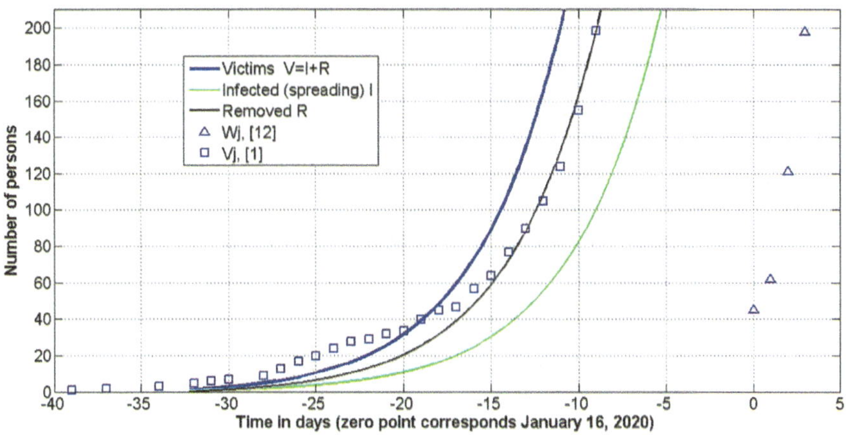

Fig. 6.4 Mainland China, prediction 2. Results of calculations and verification for the early stage of the epidemic. Numbers of infected I (green line), removed R (black line) and victims $V = I + R$ (blue line) versus time in days (zero point corresponds to January 16, 2020); "triangles" show cumulative number of confirmed cases W_j, reported by NHC [2]; the "squares" correspond to the cumulative number of cases, summarized with the use of daily distribution available in [6, Fig. 1]

6.2 The Republic of Korea

Here, we consider the development of the COVID-19 epidemic in the Republic of Korea with use the official WHO daily data ([1], Tables 1.1 and 6.3) and the approach presented in Chaps. 4 and 5.

The first prediction uses V_j data from the period of time February 17–March 12, 2020, available in Table 1.1. The results of calculations are presented in Table 6.2 (prediction 1) and in Fig. 6.5. This prediction was very optimistic, since the final number of cases was estimated as $V_\infty \approx 8117$ and the stop of local transmission was estimated after March 20, 2020 ($t_{\text{final}} \approx 26$), provided that existing in the period February 17–March 12 quarantine measures and patient isolation rates will continue.

All the parameters in SIR model are supposed to be constant. If new infected persons are still coming in the country and spread the infection, the value N is not constant. This fact can reduce the accuracy of prediction. If these "new imported" and connected with them cases can be calculated and extracted from the total number, the accuracy of the prediction could be improved.

The estimation of the average time of spreading the infection in South Korea looks incredible in this prediction ($\tau = 1/\rho = 0.1796$); i.e., this value was approximately 4.3 h. By comparison, in mainland China it was approximately 2.5 days (see prediction 2 in Table 6.1). That is why the $V(t)$ and $R(t)$ curves are very close in Fig. 6.5 and the maximal number of persons spreading the infection I

Table 6.2 Calculated optimal values of SIR model parameters for the COVID-19 epidemic in South Korea (first wave)

Number of prediction/ optimal values of parameters	Prediction 1 $n = 25$ Period taken for calculations T_c: February 17–March 12, 2020	Prediction 2 $n = 21$ Period taken for calculations T_c: April 9–29, 2020
N	68,037.824	13,679.54688
v	63,900.7878956716	6892.82635608829
α	8.71198727144046e-05	1.91761713114e-05
t_1^*	−9.19521866246193	−55.5257861059147
ρ	5.567	0.132178019
$1/\rho$	0.1796	7.5655544
r	0.9964	0.99686044433
S_∞, Eq. (4.9)	59,921	2837
V_∞, Eq. (4.16)	8117	10,842
t_{final}, Eq. (4.17)	26	127.8
Final day of the first wave	March 20, 2020	June 29, 2020
V_f, taken from [1]	8799	12,800
ε, Eq. (4.20)	7.8%	15.3%

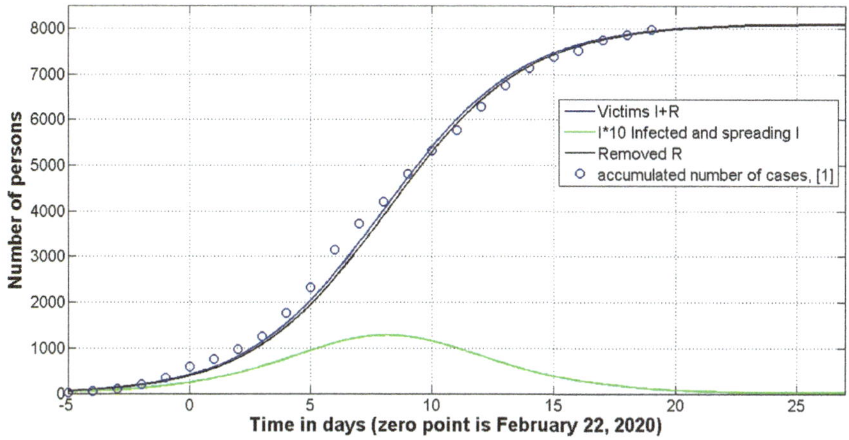

Fig. 6.5 South Korea, the first epidemic wave, prediction 1. Results of calculations. Numbers of infected I multiplied by 10 (the green line), removed R (the black line) and the number of victims $V = I + R$ (the blue line); the "circles" correspond to the confirmed accumulated number of cases in the Republic of Korea [1] shown in Table 1.1

(t) was less than 130 (to make the corresponding curve more visible in Fig. 6.5, this value is multiplied by 10). Rapid isolation allows patients to start treatment more quickly and reduce mortality. According to the WHO report [1] (March 10, 2020), the mortality rate in South Korea was 0.77%. By comparison, it was 6.22% in Italy. Such a huge difference in mortality rate can be explained by not so rapid isolation in Italy. It may also be that a more virulent coronavirus strain was spreading in Italy.

Usually, the number of cases during the initial period of an epidemic outbreak is not reliable, since many infected people are not detected. To avoid their influence on the results, only V_j values for the period April 9–29, 2020 ($47 \leq t_j \leq 67$; $n = 21$; $F_C(1, n-2) = 15.2$; see Table 6.3) were used to calculate the epidemic characteristics for South Korea and some other countries [75, 77]. Since during the quarantine, the international people exchange is quite limited, we can apply the SIR model for every country assuming its parameters to be constant (but different for every country) during the fixed period of time. The results of calculations for the Republic of Korea are shown in Table 6.2 (prediction 2). The presented second prediction for South Korea is more pessimistic in comparison with the first one. In particular, the epidemic stop was expected after June 29, 2020 (see Table 6.2), but the expected number of victims is only 25% higher for second prediction. It looks strange that the epidemic durations differ more than 5 times (see Table 6.2). Even larger are differences in the average time of spreading infection $1/\rho$. For example, the first prediction yielded incredible low value of 4.3 h, and the second one yields more realistic figure—7.6 days. Later, we will discuss possible reasons of these differences.

The SIR curves (second prediction) and markers representing the V_j values taken for calculations ("circles"), comparisons ("triangles"), and verifications of

Table 6.3 Official cumulative numbers of confirmed cases in the Republic of Korea, Germany, UK, USA and the world used for calculations and verifications of predictions [1, 77]

Day in April and May, 2020	Time in days t_j	Number of cases in South Korea	Number of cases in Germany	Number of cases in UK	Number of cases in USA	Global number of cases
9	47	10,450	113,525	65,081	425,889	1,521,252
10	48	10,480	117,658	70,276	461,275	1,610,909
11	49	10,512	120,479	78,995	492,881	1,696,588
12	50	10,537	123,016	84,283	524,514	1,773,084
13	51	10,564	125,098	88,625	553,822	1,844,863
14	52	10,591	127,584	93,877	578,268	1,914,916
15	53	10,613	130,450	98,480	604,070	1,991,562
16	54	10,635	133,830	103,097	632,781	2,074,529
17	55	10,653	137,439	108,696	665,330	2,160,207
18	56	10,661	139,897	114,221	695,353	2,241,359
19	57	10,674	141,672	120,071	723,605	2,314,621
20	58	10,683	143,457	124,747	751,273	2,397,216
21	59	10,694	145,694	129,048	776,907	2,471,136
22	60	10,702	148,046	133,499	800,926	2,544,792
23	61	10,708	150,383	138,082	830,053	2,626,321
24	62	10,718	152,438	143,468	860,772	2,719,897
25	63	10,728	154,175	148,381	899,281	2,804,796
26	64	10,738	155,193	152,844	931,698	2,878,196
27	65	10,752	156,337	157,153	960,916	2,954,222
28	66	10,761	157,641	161,149	983,457	3,018,952
29	67	10,765	159,119	165,225	1,003,974	3,090,445
30	68	10,774	159,119	171,257	1,035,353	3,175,207
1	69	10,780	161,703	177,458	1,067,127	3,267,184
2	70	10,793	162,496	182,264	1,093,880	3,349,786
3	71	10,801	163,175	186,603	1,125,719	3,435,894
4	72	10,804	163,860	190,588	1,154,985	3,517,345
5	73	10,806	164,897	194,994	1,171,185	3,588,773
7	75	10,822	167,300	206,719	1,215,571	3,759,967

calculations (“stars”) are shown in Figs. 6.6 and 6.7 by magenta color for South Korea. Other countries and regions are also represented in these figures with the use of different colors. Corresponding optimal values of SIR parameters will be presented later in this chapter.

Figures 6.6 and 6.7 demonstrate that the confirmed number of cases in the Republic of Korea deviated not very significantly from the theoretical estimation. Magenta “triangles” approach magenta lines rather fast and coincide after March 1, 2020. It means that infected persons were detected and isolated faster in comparison

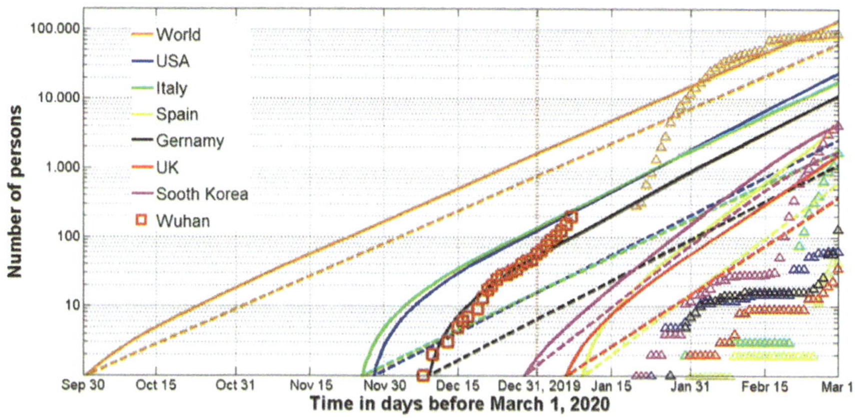

Fig. 6.6 SIR curves (first epidemic waves, lines) and accumulated number of cases (markers) versus time before March 1, 2020. Numbers of infected and spreading I (dashed) and victims $V = I + R$ (solid); "triangles" represent the V_j values taken for comparisons. Colors correspond to the country or region: world—brown; USA—blue; Italy—green; Spain—yellow; UK—red and South Korea—magenta. The number of laboratory confirmed cases in Wuhan, China, is shown by brown "squares". These values were calculated in [69] with the use of information from [6]

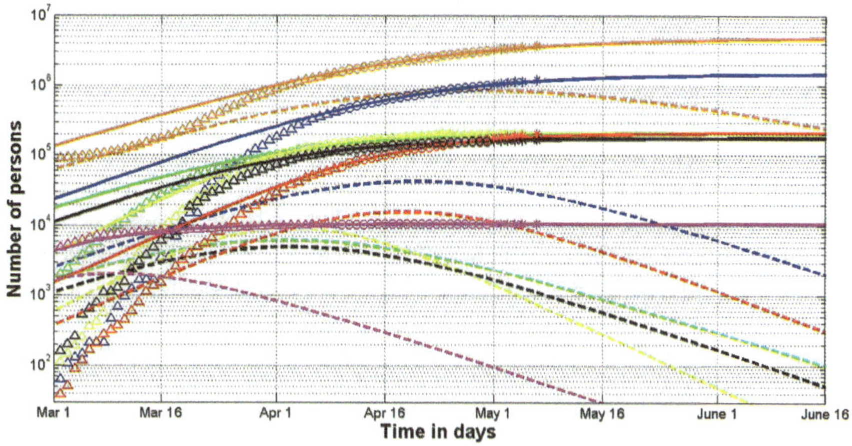

Fig. 6.7 SIR curves (first epidemic waves, lines) and accumulated number of cases (markers) versus time after March 1, 2020. Numbers of infected and spreading I (dashed) and victims $V = I + R$ (solid); markers represent the V_j values taken for calculations ("circles"), comparisons ("triangles") and verifications of calculations ("stars"). Colors correspond to the country or region: world—brown; USA—blue; Italy—green; Spain—yellow; UK—red and South Korea—magenta

with other countries. This is probably the main reason for Korea's success in fighting the COVID-19 pandemic. Figure 6.7 shows much lower number of cases in comparison with other countries with similar number of citizens.

As noted in the previous chapters, the applied approach can describe only the first wave of the epidemic and does not take into account imported cases, so to assess its accuracy ε, we can use information on the accumulated number of cases V_f recorded in the day, which corresponds to t_{final}, and formula (4.20). The corresponding values are shown in Table 6.2 and demonstrate that the accuracy of both predictions are rather good ($\varepsilon = 7.8\%$ and 15.3%), but the discrepancy of the second prediction for China was lower ($\varepsilon = 3.8\%$).

As of September 26, 2020, the number of cases in South Korea was 23,611 [52]. It means that for 4.5 months of observation of the epidemic (after the publication of the second prediction in [75]), the revealed exceeding the forecast level was only 1.18 times. Thus, both predictions for the final sizes of the COVID-19 epidemic in Korea were not bad, especially if we take into account that the used SIR model supposes the constant values of its parameters (in particular, zero number of imported cases). Later, we will discuss how to take into account changes in the SIR model parameters and increase the accuracy of predictions.

6.3 Italy

In this section, we consider the first wave of the COVID-19 epidemic in Italy. The results of comparison with the second prediction for mainland China and several SIR simulations will be presented.

Dramatic situation in Italy in March–June 2020 forced to pay much attention to the pandemic dynamics in this country. The first study is already described in Chap. 3. It concerns the comparison of the number of cases in Italy with the number of cases W_j in China available in Table 1.2. Due to the fact that the Chinese data were incomplete by February 10, 2020 (see [2, 42, 43, 67]), there was a need for a new comparison of the dynamics in these two countries. After the publication of the results of the second forecast for China [69], it became possible to compare the resulting $V(t)$ curve with the number of cases registered in Italy immediately after the outbreak (available in Table 3.1).

Since the epidemics in these countries started at different times, it was necessary to make time synchronization (see [50]). Let us suppose that the epidemic outbreak in Italy happened on March 22, 2020 (the cumulative number of cases in this country was 76; see Table 3.1). The time moment t_s corresponding to the number of victims $V = 76$ was calculated with the optimal values of parameters for the second prediction for mainland China available in Table 6.1. The result $t_s = -15.82545$ means that all the corresponding time moments for the data set for Italy have to be shifted by the value 52.82545, since March 22 corresponds to the time moment $t_j = 37$ (see Table 1.2, where zero t_j value corresponds to January 16, 2020). The "stars" in Fig. 6.8 represent the accumulated number of confirmed cases in Italy with the corresponding time shift. The line shows the $V = I + R$ values corresponding to the second prediction for China (see Table 6.1 and Fig. 6.4).

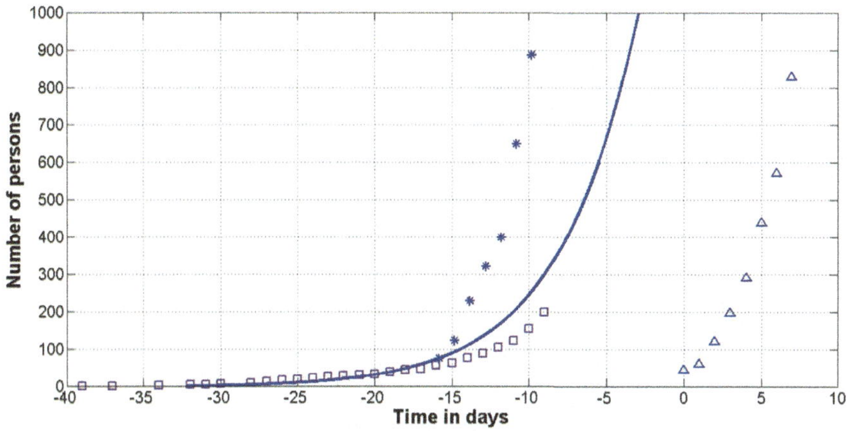

Fig. 6.8 Comparison of the early stage of the epidemic dynamics in Italy with the corresponding SIR curve for mainland China. Number of victims $V = I + R$ (blue line) versus time in days (zero point is January 16, 2020) estimated in [50] for the mainland China; "triangles" represent the official data set for China (see Table 1.2); "squares"—the cumulative number of cases in Wuhan calculated with the information presented in [6] (see Table 1.2); "stars"—to the points corresponding to the epidemic in Italy ([1], Table 3.1) with the corresponding time shift

Figure 6.8 demonstrates that the epidemic in Italy developed much more rapid than it was in China. Thus, the application of SIR model allowed predicting the higher final number of cases in Italy already on March 3, 2020, when the preprint [50] was submitted.

The first SIR simulation for Italy was published in [38] with the use of data presented in Tables 2.1 or 6.5. Usually, the number of cases during the initial period of an epidemic outbreak is not reliable, since many cases are not detected (see, e.g., Fig. 6.6). To avoid their influence on the results, only V_j values for the period March 5–22, 2020 ($12 \le t_j \le 29$), were used for calculations (see "circles" in Fig. 6.9). Other points were used only for comparison ("triangles") and verification of predictions ("stars"). The use of data corresponding to the initial stage of the epidemic ("circles" and "triangles" together) did not yield a stable reliable prediction. The calculated optimal values of parameters are presented in Table 6.4 (prediction 1).

According to this prediction, in Italy more people were expected to be infected ($V_\infty \approx 111,548$) in comparison with mainland China, where the saturation level $V_\infty \approx 81,257$ was predicted in [69] on March 4, 2020 (see Sect. 6.1). According to this estimation, we can expect that local transmission of the epidemic in Italy was expected to stop only after May 12, 2020, provided that quarantine measures of period T_c would continue until mid-May.

It was also possible to calculate the value of parameter $\rho = \nu\alpha = 1.6564$ and the inverse value $1/\rho = 0.6037$. Thus, the average time of spreading the infection in Italy was estimated as 14.5 h. By comparison, in South Korea it was estimated to be

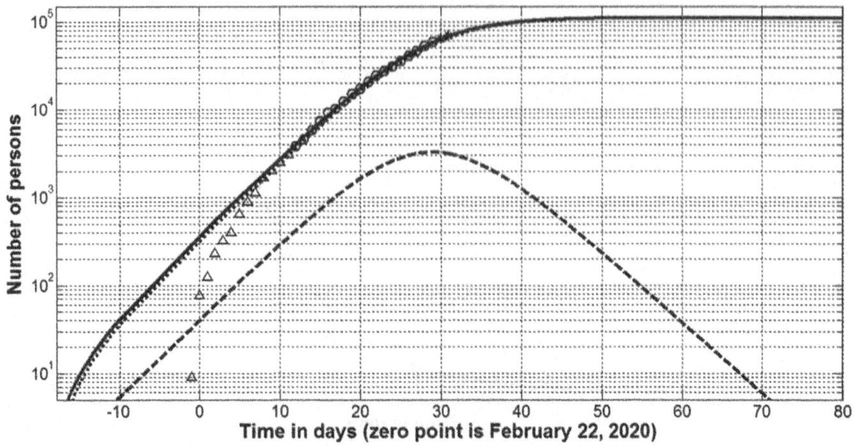

Fig. 6.9 First wave of pandemic in Italy, prediction 1. Results of SIR simulations. Numbers of infected I (dashed line), removed R (dotted line) and the number of victims $V = I + R$ (solid line); "circles" correspond to the confirmed accumulated number of cases taken for calculations; "triangles" correspond to the cases during initial stage of the epidemic; "stars"—two data points used only for a verification of the prediction

approximately 4.3 h and in mainland China—2.5 days (see Sects. 6.1 and 6.2). By mid-April 2020, there were more than thousand people spreading the infection in Italy (see dashed line in Fig. 6.9). The calculated value t_1^*, solid line and "triangles" in Fig. 6.9 demonstrate that the first cases of COVID-19 have not been identified in Italy and sick people overspread it rather long time.

After the saturation level $V_\infty \approx 111,548$ of the first prediction was exceeded, 3 new predictions for Italy have been performed with the use of the official WHO data, listed in Tables 6.5 and 6.6. To investigate the influence of the number of observations n and the period used for calculations $t_{j1} \le T_c \le t_{j2}$, different values of initial t_{j1} and final t_{j2} moments of time were used. The results of calculations are shown in Table 6.4 (predictions 2, 3 and 4). The SIR curves corresponding to the prediction 4 are shown in Fig. 6.10. Markers represent the V_j values taken for calculations ("circles"), for a comparison ("triangles") and a verification of predictions ("star").

Figure 6.10 illustrates that the real epidemic outbreak could start even in January 2020. Probably, first cases of the COVID-19 infection were not identified and sick people were not isolated. It could be a reason of the high saturation level of the epidemic (approximately 148,000–153,000 according to the predictions 3 and 4). The finish of the first wave of the epidemic in Italy was predicted to happen on May 25–31, 2020. The average time of spreading infection $1/\rho$ could be estimated as 0.7–0.75 days (according to predictions 3 and 4, see Table 6.4).

The discrepancies ε shown in the last row of the Table 6.4 are much higher then we have obtained for China (second prediction) and South Korea (see last row in Tables 6.1 and 6.2). Observations of the epidemic in Italy showed that the

Table 6.4 Predictions 1–4 for the first wave of the COVID-19 epidemic in Italy

Number of prediction/optimal values of parameters, characteristics	Prediction 1 Period taken for calculations T_c: March 5–22, 2020	Prediction 2 Period taken for calculations T_c: March 10–30, 2020	Prediction 3 Period taken for calculations T_c: March 10–April 5, 2020	Prediction 4 Period taken for calculations T_c: March 16–April 5, 2020
N	534,656	669,312	723,763.2	738,240
v	476,712.7441113	598,876.49226424	646,709.93100256	658,709.4638592
α	3.47467962e-06	2.589545071e-06	2.2076760085e-06	2.0049693740e-06
t_1^*	−18.3797687613	−22.039683744	−25.171859952631	−28.810667619253
ρ	1.656424058815	1.5508176688225	1.4277259991363	1.3206923014088
$1/\rho$	0.603710139730	0.6448211289462	0.7004145057279	0.7571786395160
t_{j1}	12	17	17	23
t_{j2}	29	37	43	43
n	18	21	27	21
r	0.999378228510	0.999725363341	0.9992432785563	0.9992024895
F, Eq. (2.9)	12,854.46446586	34,576.908427910	16,499.880328838	11,897.820763752
$F_C(1, n-2)$	16.2	15.2	13.9	15.2
$F/F_C(1, n-2)$	793.4854608556	2274.7966070993	1187.0417502761	782.75136603635
S_∞, Eq. (4.9)	423,108	533,558	575,323	585,100
V_∞, Eq. (4.16)	111,548	135,754	148,441	153,140
t_{final}, Eq. (4.17)	79.5	87.8	93.9	98.8
Final day of the first wave	May 12, 2020	May 20, 2020	May 26, 2020	May 31, 2020
V_f, taken from [1]	221,216	227,364	230,555	233,019
ε, Eq. (4.20)	49.6%	40.3%	35.6%	34.3%

Optimal values of parameters and other SIR model characteristics

Table 6.5 Official cumulative numbers of confirmed cases in Italy, Spain, Germany, France, Austria and Moldova after the COVID-19 pandemic outbreak in Europe [1, 74]

Day in February and March, 2020	Time moments in days t_j	Number of cases in Italy	Number of cases in Spain	Number of cases in Germany	Number of cases in France	Number of cases in Austria	Number of cases in the Republic of Moldova
21	−1	9	2	16	12	0	0
22	0	76	2	16	12	0	0
23	1	124	2	16	12	0	0
24	2	229	2	16	12	0	0
25	3	322	2	18	12	2	0
26	4	400	12	21	18	2	0
27	5	650	25	26	38	4	0
28	6	888	32	57	57	5	0
29	7	1128	45	57	100	10	0
1	8	1689	45	129	100	10	0
2	9	2036	114	157	191	18	0
3	10	2502	151	196	212	24	0
4	11	3089	198	262	282	37	0
5	12	3858	257	534	420	47	0
6	13	4636	374	639	613	66	0
7	14	5883	430	795	706	104	1
8	15	7375	589	1112	1116	112	1
9	16	9172	1024	1139	1402	131	1
10	17	10,149	1639	1296	1774	182	3
11	18	12,462	2140	1567	2269	302	4
12	19	15,113	2965	2369	2860	361	4
13	20	17,660	4231	3062	3640	504	8
14	21	21,157	5753	3795	4469	800	12
15	22	24,747	7753	4838	5380	959	23
16	23	27,980	9191	6012	6573	1132	29
17	24	31,506	11,178	7156	7652	1332	30
18	25	35,713	13,716	8198	9043	1646	36
19	26	41,035	17,147	10,999	10,877	1843	49
20	27	47,021	19,980	18,323	12,475	2649	66
21	28	53,578	24,926	21,463	14,296	3024	80
22	29	59,138	28,572	24,774	15,821	3631	94
23	30	63,927	33,089	29,212	19,615	4486	109
24	31	69,176	39,673	31,554	22,025	5282	125
25	32	74,386	47,610	36,508	24,920	5888	149
26	33	80,539	56,188	42,288	28,786	7029	177
27	34	86,498	64,059	48,582	32,542	7697	199

Table 6.6 Official cumulative numbers of confirmed cases in Italy, Spain, Germany, France, Austria and Moldova in March and April 2020 [1, 74]

Day in March and April, 2020	Time moments in days t_j	Number of cases in Italy	Number of cases in Spain	Number of cases in Germany	Number of cases in France	Number of cases in Austria	Number of cases in the Republic of Moldova
28	35	92,472	72,248	52,547	37,145	8291	231
29	36	97,689	78,797	57,298	39,642	8813	263
30	37	101,739	85,195	61,913	43,977	9618	298
31	38	105,792	94,417	67,366	51,477	10,182	353
1	39	110,574	102,136	73,522	56,261	10,711	423
2	40	115,242	110,238	79,696	58,327	11,129	591
3	41	119,827	117,710	85,778	63,536	11,525	591
4	42	124,632	124,736	91,714	67,757	11,766	752
5	43	128,948	130,759	95,391	69,607	11,983	864
6	44	132,547	135,032	99,225	73,488	12,297	965
7	45	135,586	140,510	103,228	77,226	12,640	1056
8	46	139,422	146,690	108,202	81,095	12,969	1174
9	47	143,626	152,446	113,525	85,351	13,248	1289
10	48	147,577	157,022	117,658	89,683	13,560	1438
11	49	152,271	161,852	120,479	92,787	13,807	1560
12	50	156,363	166,019	123,016	94,382	13,937	1662
13	51	159,516	169,496	125,098	97,050	14,043	1712
14	52	162,488	172,541	127,584	102,533	14,234	1934
15	53	165,155	177,633	130,450	105,155	14,370	2049
16	54	168,941	182,816	133,830	107,778	14,448	2154
17	55	172,434	188,068	137,439	108,163	14,603	2264
18	56	175,925	191,726	139,897	110,721	14,662	2351
19	57	178,972	195,944	141,672	111,463	14,710	2472
20	58	181,228	200,210	143,457	113,513	14,783	2548

registered number of cases in this country quickly exceeded the saturation levels V_∞ presented in Table 6.4. This fact motivated to make new calculations using more recent data from Table 6.6. The results are given in Table 6.7 and in Fig. 6.11.

The value V_f for the prediction 6 was interpolated with the use figures for accumulated number of cases in Italy 258,136 (on August 22, 2020) and weekly increase 4698 available in WHO report [82]. Unfortunately, WHO stopped to present daily information after August 16, 2020. The SIR curves corresponding to prediction 6 are presented in Fig. 6.11.

It can be seen that the previous predictions for Italy were more optimistic. Fresh data sets have showed that the final number of cases in Italy could reach 226,000

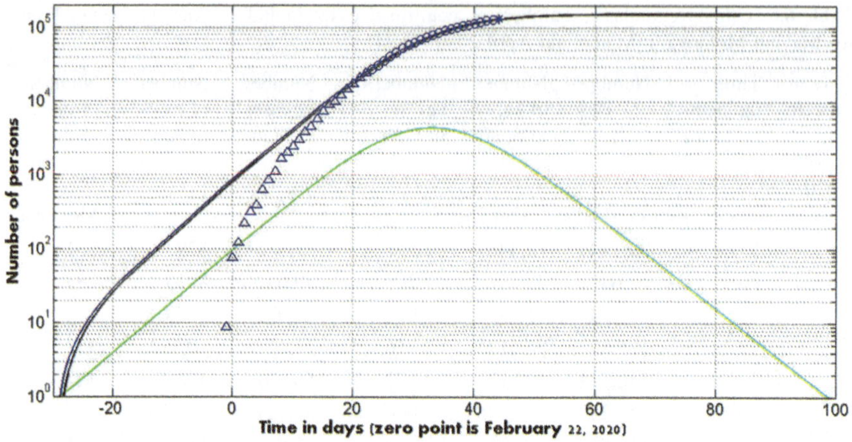

Fig. 6.10 First wave of COVID-19 epidemic Italy, prediction 4: SIR curves (lines) and accumulated number of cases (markers) versus time. Numbers of infected I (green), removed R (black) and the number of victims $V = I + R$ (blue line); "circles" correspond to the confirmed accumulated number of cases taken for calculations; "triangles" correspond to the cases during initial stage of the epidemic; "star"—a data point used only for a verification of the prediction

Table 6.7 Predictions 5 and 6 for the first epidemic wave in Italy

Number of prediction/optimal values of parameters, characteristics	Prediction 5 $n = 14$ Period taken for calculations T_c: March 28–April 10, 2020	Prediction 6 $n = 21$ Period taken for calculations T_c: March 29–April 18, 2020
N	984,556.8	1,125,589.888
v	882,151.283536331	1,008,518.7886
α	9.4016573087e-07	6.759461602e-07
t_1^*	−67.283729213432	−88.26304932
ρ	0.8293684062279	0.681704402658
$1/\rho$	1.20573679017765	1.46691145913
r	0.9996573802724	0.99969602932
F, Eq. (2.9)	17,503.1269829749	31,238.7652767
$F/F_C(1, n-2)$	941.028332418004	2055.181926
S_∞, Eq. (4.9)	787,098	899,854
V_∞, Eq. (4.16)	197,459	225,736
t_{final}, Eq. (4.17)	148.5	177.8
Final day of the first wave	July 20, 2020	August 18, 2020
V_f	244,624 [1]	255,451 [82]
ε, Eq. (4.20)	19.3%	11.6%

Optimal values of parameters and other SIR model characteristics

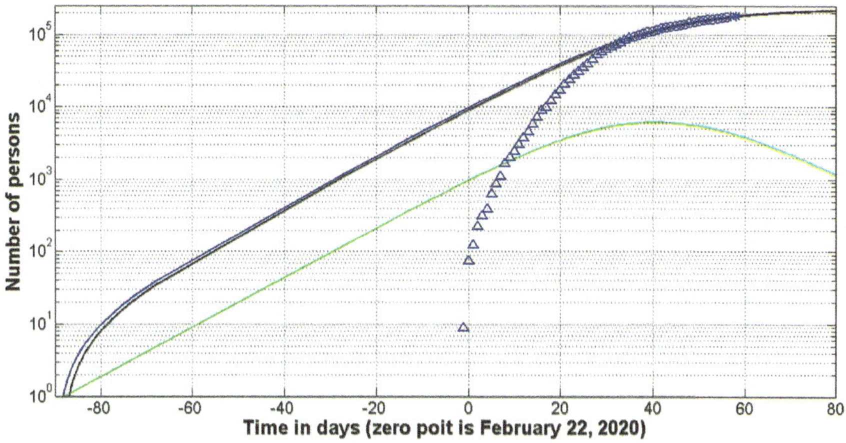

Fig. 6.11 First wave of pandemic in Italy, prediction 6: SIR curves (lines) and accumulated number of cases (markers) versus time. Numbers of infected I (green), removed R (black) and victims $V = I + R$ (blue line); "circles" correspond to the confirmed accumulated number of cases taken for calculations; "triangles" correspond to the cases during initial stage of the epidemic; "stars"—data points used only for a verification of the prediction

and the first wave of the epidemic can stop only after August 18, 2020 (see Table 6.7, prediction 6). New estimations of the relative error ε are much lower than presented in Table 6.4. The reliability of prediction 6 is very high, because for four months of monitoring the epidemic in Italy (this forecast was posted on April 22, 2020 [83]) the real number of cases exceeded the estimated saturation level $V_\infty = 225,736$ by only 11.6%. The higher error of the previous calculations can be explained by the fact that they used data from earlier periods, when not all cases of the disease were detected. Figures 6.9, 6.10 and 6.11 show large differences between the registered number of cases ("triangles") and theoretical $V = I + R$ (curves).

Figures 6.8 and 6.11 and Tables 6.4 and 6.7 illustrate that the estimations of t_1^* values are very different. The probable beginning of the epidemic in Italy was postponed to an earlier time with the use of more recent data used for calculations. In particular, according to the prediction 6, the first COVID-19 cases could happen even after November 27, 2019. This result correlates with the information from Giuseppe Remuzzi, director of the Mario Negri Institute for Pharmacological Research, that "virus was circulating before we were aware of the outbreak in China" [84].

It must be noted that the highest level of the $F/F_C(1, n-2)$ ratio was achieved in prediction 2 with the error $\varepsilon = 40.3\%$ (compare corresponding values in Tables 6.4 and 6.7). This fact allows us to conclude that estimating the quality of forecasts should be based not only on the value of this ratio. The estimate ε is the most adequate characteristic of the quality of calculations, but it can be calculated only after the point in time t_{final} that can be expected in months (as we see in the example of prediction 6). To more quickly assess the quality of the forecast, it is necessary to constantly monitor the real epidemic dynamics and refine the calculations using more recent data if there are large discrepancies between the actually recorded number of cases and the theoretical $V = I + R$ curves.

6.4 Austria

In this section, we consider the first wave of the COVID-19 epidemic in Austria and present several SIR simulations.

The author is very grateful to his Austrian friend and co-author Gerhard Demelmair who did a tremendous job of collecting and systematizing statistical information (it is presented in numerous tables), so a lot of attention was paid to SIR simulations of the epidemic in Austria [71, 74, 75, 77, 78]. The corresponding data sets are presented in Tables 6.8 and 6.9. The results of two first SIR simulations with the use of different periods T_c taken for calculations are shown in Table 6.10.

To calculate the first prediction, we took only the data points for the period March 6–18 ($13 \leq t_j \leq 25$). The incredible small figure for the saturation level $V_\infty \approx 3589$ was obtained. The calculations also gave the value $t_{\text{final}} \approx 47.1$. According to this estimation, the local transmission in Austria could stop after April 10, 2020. The corresponding SIR curves are shown in Fig. 6.12.

The real number of cases deviated very fast from the solid line shown in Fig. 6.12. After April 10, it was possible to calculate the error according to discrepancy according to the formula (4.20). The value $\varepsilon = 73.5\%$ turned to be too large. We have used more fresh V_j data (for the period March 14–29, 2020) to calculate second prediction. The results are also presented in Table 6.10. Much higher figures for the saturation level $V_\infty \approx 14,251$ and epidemic duration ($t_{\text{final}} \approx 67.3$) were obtained. The value of ε was only 7.8%. These results demonstrate once more that the numbers of COVID-19 cases are not reliable immediately after the epidemic outbreak, since many infected persons remain not detected.

It must be noted that the correlation coefficient for the first prediction is slightly higher (see Table 6.10). Therefore, the value ε is better estimation of the reliability of prediction. If the real number of cases is much higher than the predicted saturation level V_∞, there is no need to wait until predicted moment of time t_{final}. In such situations, new estimations with the use of fresher data sets are necessary. We have done this with the use of also some V_j values from Table 6.11. The results of such calculations are shown in Tables 6.10 and 6.12.

The obtained values of ε are rather low, and for predictions 2 and 5 they are very close (see Tables 6.10 and 6.12). Also close are the estimations for parameters N, v, α, t_1^* and for the final day of the first wave of the epidemic in Austria for predictions 2–5. The average time of spreading infection $1/\rho$ was estimated as 0.44–0.77 days and is much lower than in Italy (see Table 6.7), but higher than for the first

Table 6.8 Official cumulative numbers of confirmed cases in Austria

Day in February and March, 2020	Accumulated number of cases V_j [1]	Time moments in days t_j
22	0	0
23	0	1
24	0	2
25	2	3
26	2	4
27	2	5
28	5	6
29	10	7
1	10	8
2	18	9
3	18	10
4	37	11
5	47	12
6	66	13
7	104	14
8	112	15
9	131	16
10	182	17
11	302	18
12	361	19
13	504	20
14	655	21
15	860	22
16	1016	23
17	1332	24
18	1646	25
19	1843	26

The information from [1]. The corresponding time moments t_j and the accumulated confirmed numbers of cases V_j in Austria

prediction for South Korea. The SIR curves corresponding prediction 3 (with the highest value of $F/F_C(1, n-2) = 981.01748$) are shown in Fig. 6.13.

Despite the highest value of the $F/F_C(1, n-2)$ ratio, the number of cases registered at the beginning of the epidemic V_j ("triangles" in Fig. 6.13) is not less than the theoretical curve V calculated using later data ("circles" in Fig. 6.13), and some of them even exceed the theoretical estimation. This result casts doubt on the adequacy of prediction 3, as the experience of calculating the dynamics of the COVID-19 pandemic in many countries shows that a large number of patients remain undetected at the beginning of epidemics (see, e.g., Sects. 6.1–6.3). Prediction 6 does not have such a shortcoming. Its characteristics are shown in Table 6.12, and the corresponding curves are presented in Fig. 6.14.

Table 6.9 Official cumulative numbers of confirmed cases in Italy, Austria and Ukraine used for calculations, comparisons and verifications of predictions [1, 71]

Day in March and April, 2020	Time moments in days t_j	Number of cases in Italy V_J	Number of cases in Austria V_{Aj}	Number of cases in Ukraine V_{Uj}
5	12	3858	47	1
6	13	4636	66	1
7	14	5883	104	1
8	15	7375	112	1
9	16	9172	131	1
10	17	10,149	182	1
11	18	12,462	302	1
12	19	15,113	361	3
13	20	17,660	504	3
14	21	21,157	800	3
15	22	24,747	959	3
16	23	27,980	1132	7
17	24	31,506	1332	14
18	25	35,713	1646	16
19	26	41,035	1843	16
20	27	47,021	2649	26
21	28	53,578	3024	47
22	29	59,138	3631	47
23	30	63,927	4486	84
24	31	69,176	5282	113
25	32	74,386	5888	156
26	33	80,539	7029	218
27	34	86,498	7697	311
28	35	92,472	8291	418
29	36	97,689	8813	480
30	37	101,739	9618	549
31	38	105,792	10,182	669
1	39	110,574	10,711	804
2	40	115,242	11,129	987
3	41	119,827	11,525	1096
4	42	124,632	11,766	1251
5	43	128,948	11,983	1319
6	44	132,547	12,297	1462

It turned out that this forecast gives the smallest values of error $\varepsilon = 6.6\%$; therefore, it can be considered as one of the best predictions based on classical SIR model. As of September 26, 2020 (see WHO weekly update report [52]), the number of cases in Austria 42,940 exceeded 2.8 times the saturation level

Table 6.10 Predictions 1–3 for the first wave of the COVID-19 epidemic in Austria

Number of prediction/optimal values of parameters, characteristics	Prediction 1 $n = 13$ Period taken for calculations T_c: March 6–18, 2020	Prediction 2 $n = 16$ Period taken for calculations T_c: March 14–29, 2020	Prediction 3 $n = 23$ Period taken for calculations T_c: March 14–April 5, 2020
N	12,529.6	71,099.22048	65,613.696
v	10,636.672136192	63,712.2688217289	58,821.6712255282
α	0.000176292342229	3.190277719e-05	3.63707317842e-05
t_1^*	6.60315754396316	2.38559451409778	3.38412182748648
ρ	5.567	2.03259831679520	2.13938722724589
$1/\rho$	0.1796	0.491981121767681	0.467423562814918
r	0.998805	0.99806	0.999267711104992
S_∞, Eq. (4.9)	8940	56,848	52,508
V_∞, Eq. (4.16)	3589	14,251	13,106
t_{final}, Eq. (4.17)	47.1	67.3	64.5
Final day of the first wave	April 10, 2020	April 30, 2020	April 27, 2020
V_f, taken from [1]	13,560	15,457	15,256
ε, Eq. (4.20)	73.5%	7.8%	14.1%

Optimal values of parameters and other SIR model characteristics

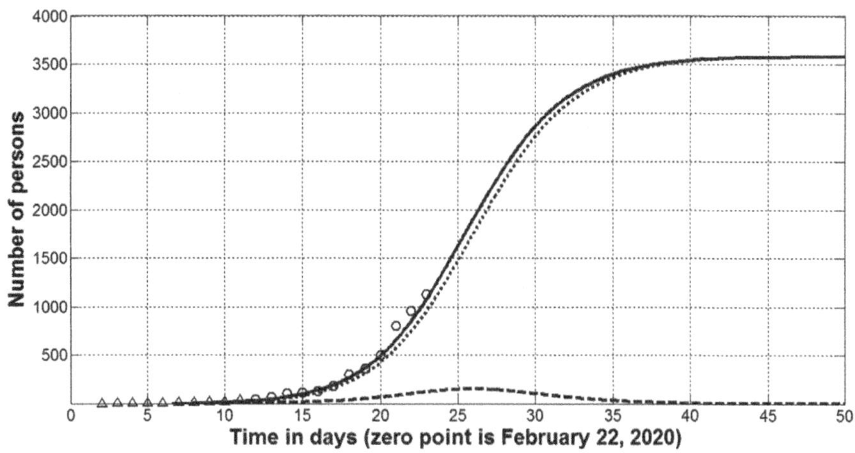

Fig. 6.12 Austria, first epidemic wave, prediction 1. Results of calculations. Numbers of infected I multiplied by 10 (dashed line), removed R (dotted line) and the number of victims $V = I + R$ (solid line); the "circles" correspond to the confirmed accumulated number of cases in Austria taken for calculations; "triangles" show the cases before March 6, 2020

Table 6.11 Official cumulative numbers of confirmed cases in Ukraine and Austria used for calculations, verifications of predictions and estimations of the pandemic dynamics after quarantines weakening [1, 78]

Day in April and May, 2020	Time in days t_j	Number of cases in Ukraine	Number of cases in Austria	Day in May and June, 2020	Time in days t_j	Number of cases in Ukraine	Number of cases in Austria
25	96	8617	15,134	21	122	20,148	16,332
26	97	9009	15,189	22	123	20,580	16,361
27	98	9410	15,256	23	124	20,986	16,407
28	99	9866	15,314	24	125	21,245	16,439
29	100	10,406	15,364	25	126	21,584	16,459
30	101	10,861	15,457	26	127	21,905	16,497
1	102	11,411	15,531	27	128	22,382	16,515
2	103	11,913	15,558	28	129	22,811	16,543
3	104	12,331	15,597	29	130	23,294	16,594
4	105	12,697	15,621	30	131	23,672	16,638
5	106	13,184	15,586	31	132	24,002	16,642
6	107	13,691	15,651	1	133	24,340	16,663
7	108	14,195	15,673	2	134	24,823	16,674
8	109	14,710	15,735	3	135	25,411	16,705
9	110	15,232	15,777	4	136	25,964	16,741
10	111	15,648	15,787	5	137	26,514	16,803
11	112	16,023	15,874	6	138	26,999	16,822
12	113	16,425	15,910	7	139	27,462	16,868
13	114	16,847	15,964	8	140	27,856	16,889
14	115	17,330	16,005	9	141	28,381	16,902
15	116	17,858	16,068	10	142	29,070	16,936
16	117	18,291	16,140	11	143	29,753	16,964
17	118	18,616	16,154	12	144	30,506	16,994
18	119	18,876	16,179	13	145	31,154	17,014
19	120	19,230	16,257	14	146	31,810	17,038
20	121	19,706	16,275	15	147	32,476	17,065

It should be noted some incorrectness in WHO data (the number of cases in Austria corresponding May 5 is lower than for the previous day)

$V_\infty = 15{,}108$ (see Table 6.12), predicted and posted in [83] on April 22, 2020. The reason for these discrepancies is that the theory used in this chapter describes only the first waves of the epidemic with fixed characteristics. Changes in quarantine conditions and social behavior have changed the parameters of the epidemic and made the classical SIR model unusable. Later, we will discuss how changes in the characteristics of the epidemic can be taken into account in a generalized SIR model.

Table 6.12 Predictions 4–6 for the first wave of the COVID-19 epidemic in Austria

Number of prediction/ optimal values of parameters, characteristics	Prediction 4 $n = 16$ Period taken for calculations T_c: March 21–April 5, 2020	Prediction 5 $n = 21$ Period taken for calculations T_c: March 21–April 10, 2020	Prediction 6 $n = 14$ Period taken for calculations T_c: March 28–April 10, 2020
N	64,200	71,300	75,176.032
v	57,523.2	63,884.8	67,343.8674559792
α	3.91631629710e-05	2.73374610690213e-05	1.924971386379e-05
t_1^*	4.88499029942458	−2.43641632159448	−15.8059189609884
ρ	2.25279045621570	1.74644823290221	1.29635017900866
$1/\rho$	0.443893925971 18	0.572590690729046	0.771396507049289
r	0.9994765957676	0.996416626357582	0.996465578483694
F, Eq. (2.9)	13,363.4852428387	2636.89149376656	1688.59570093583
$F/F_C(1, n-2)$	772.455794383737	173.479703537274	90.7847151040770
S_∞, Eq. (4.9)	51,319	56,995	60,068
V_∞, Eq. (4.16)	12,881	14,304	15,108
t_{final}, Eq. (4.17)	62.6	73.1	86.5
Final day of the first wave	April 25, 2020	May 5, 2020	May 18, 2020
V_f, taken from [1]	15,134	15,586	16,179
ε, Eq. (4.20)	14.9%	8.2%	6.6%

Optimal values of parameters and other SIR model characteristics

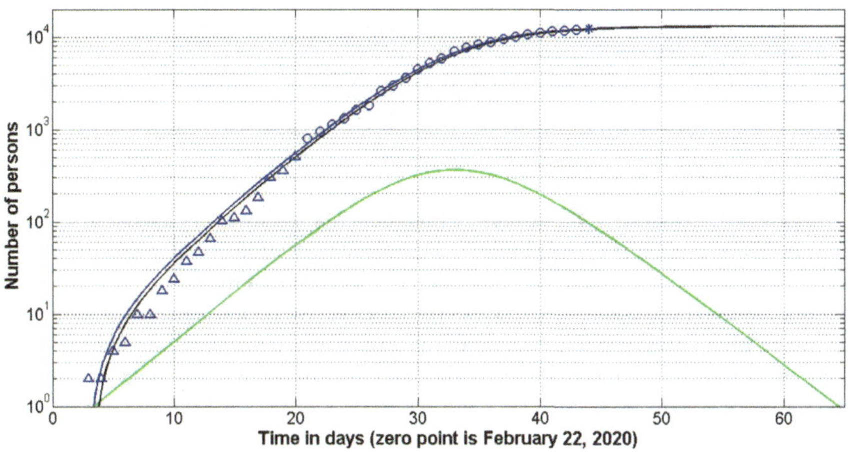

Fig. 6.13 Austria, prediction 3: SIR curves (lines) and accumulated number of cases (markers) versus time. Numbers of infected I (green), removed R (black) and the number of victims $V = I + R$ (blue line); "circles" show the cases taken for calculations; "triangles" correspond to the cases during initial stage of the epidemic; "star"—the last data point used only for a verification of the prediction

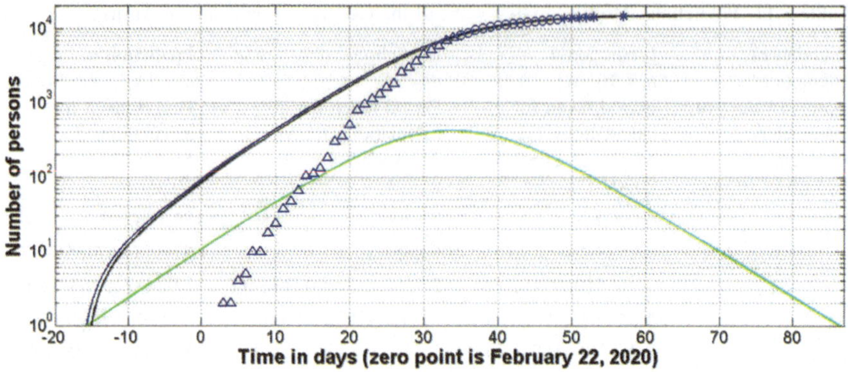

Fig. 6.14 Austria, first epidemic wave, prediction 6: SIR curves (lines) and accumulated number of cases (markers) versus time. Numbers of infected I (green), removed R (black) and the number of victims $V = I + R$ (blue line); "circles" show the cases taken for calculations; "triangles" correspond to the cases during initial stage of the epidemic; "stars"—the last data points used only for a verification of the prediction

The average time of spreading infection $1/\rho$ in Austria was estimated as 0.77 days (compare with the almost twice higher value—1.47 days—in Italy, shown in Table 6.7 for corresponding prediction 6). The proper testing, identification and isolation of infected persons in Austria could be the main reason of the relative low number of cases in this country.

In Figs. 6.15 and 6.16, prediction 6 for Austria is compared with the situation in Ukraine. It can be seen that the real numbers of cases V_j ("red stars" in Fig. 6.16) recorded in Austria after the period T_c used for the calculations ("red circles" in Fig. 6.16) correspond very well to the theoretical curve V. Minor deviations begin only after April 29, 2020.

This is in line with the fact that national lockdown in Austria was canceled on April 14 (e.g., small shops were opened). This changed the characteristics of the epidemic, and after a certain incubation period we see more cases than the classical SIR theory predicted.

Figure 6.15 illustrates that real epidemic outbreak in Austria occurred much later than in Ukraine (compare solid lines). The registered numbers of cases in Austria (red "triangles") tend rather fast to the theoretical estimation $V(t)$ (red solid line). Unfortunately, in Ukraine many COVID-19 cases were not identified long time (compare blue solid blue and "triangles" in Fig. 6.15). This situation and the delay with proper testing caused much higher number of infected people in Ukraine visible in Fig. 6.16.

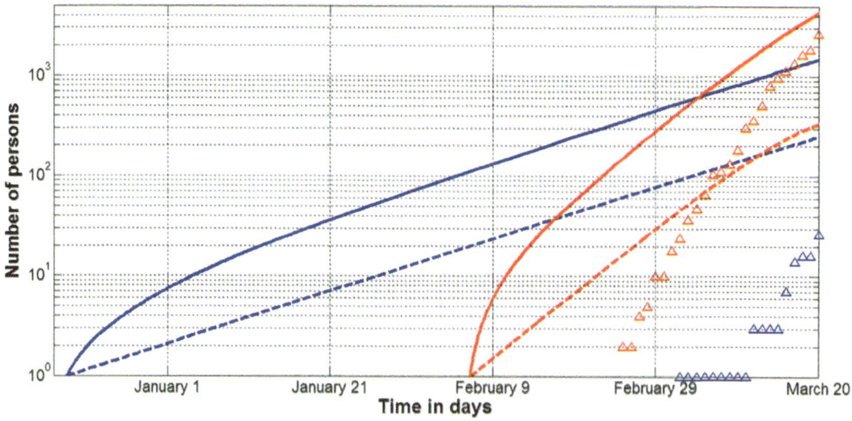

Fig. 6.15 Comparison prediction 6 for Austria with the epidemic situation in Ukraine before March 20, 2020. SIR curves (lines) and accumulated number of cases (markers) versus time before March 20, 2020. Numbers of infected and spreading I (dashed) and victims $V = I + R$ (solid); blue color corresponds to Ukraine, red—to Austria; "circles" show the cases taken for calculations; "triangles" correspond to the cases registered during initial stage of the epidemic

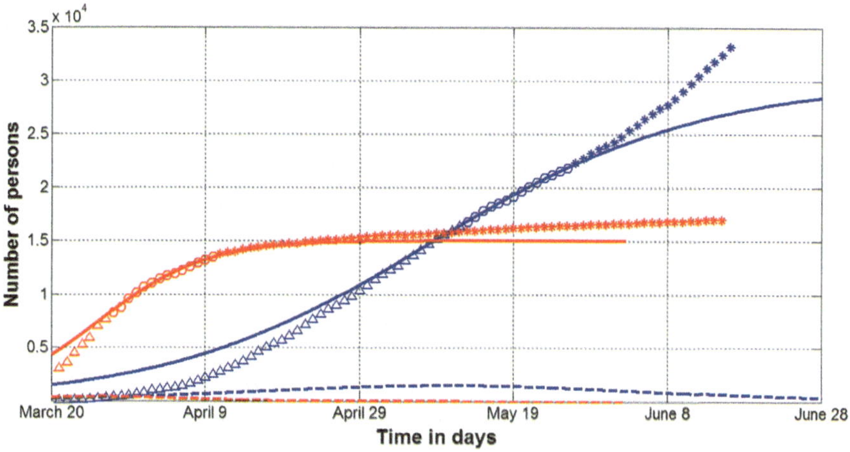

Fig. 6.16 Comparison prediction 6 for Austria with the epidemic situation in Ukraine after March 20, 2020. SIR curves (lines) and accumulated number of cases (markers) versus time after March 20, 2020. Numbers of infected and spreading I (dashed) and victims $V = I + R$ (solid); blue color corresponds to Ukraine, red—to Austria; "circles" show the cases taken for calculations; "triangles" correspond to the cases during initial stage of the epidemic; "stars"—data points used only for a verification of the predictions

6.5 Spain

In this section, we consider the first wave of the COVID-19 epidemic in Spain and present the results of three SIR simulations.

SIR simulations for the COVID-19 epidemic dynamics in Spain were performed with the use of V_j and t_j data sets presented in Tables 6.5 and 6.6 for different time periods T_c taken for calculations. The results are shown in Table 6.13 and Fig. 6.17; some of them were published in [74, 83]. The estimations for parameters N, v, α, t_1^* and for the final day of the first epidemic wave in Spain are rather different. The smallest value of ε corresponds to the prediction 2 and is rather low. Nevertheless, the highest value of $F/F_C(1, n-2)$ was obtained in prediction 3, and the highest value of the correlation coefficient r corresponds to prediction 1. Predictions 1 and 3 yield also smaller values of the saturation level V_∞ and corresponding V curves deviated faster from the V_j values registered in Spain after April 10, 2020. Therefore, prediction 2 can be treated as the most reliable and demonstrate once more that more fresh data sets yield better estimations of the COVID-19 epidemic dynamics. The average time of spreading infection $\tau = 1/\rho$ was estimated as 1.46 for this prediction. This value is very close to the value obtained in the most reliable estimation for Italy (1.47, prediction 6), but it is much higher than corresponding value in Austria (0.77, prediction 4). These results can explain much higher numbers of cases in Spain and Italy. The SIR curves corresponding to prediction 2 are shown in Fig. 6.17.

Figure 6.17 illustrates that second prediction for Spain looks too optimistic ("stars" deviate from the blue line more than for other European countries). The second prediction for this country has lower value of $F/F_C(1, n-2)$ in comparison with the predictions 1 and 3 (see Table 6.13), but for these predictions the deviations are even larger. Probably, the final size and the duration of the epidemic in Spain have to be re-estimated with the use of fresher data sets in order to decrease the value of relative error ε.

The results for Spain are compared with the situation in other countries and regions in Figs. 6.6 and 6.7. The SIR curves (second prediction) and markers representing the V_j values taken for calculations ("circles"), comparisons ("triangles") and verifications of calculations ("stars") are shown in Figs. 6.6 and 6.7 by yellow color for Spain. It can be seen that the first cases in Spain probably occurred in the beginning of January, 2020 (later than in USA, Italy, Germany, South Korea, and UK), but already in May, in terms of the number of cases, this country was second only to USA.

6.6 France

In this section, we consider the first wave of the COVID-19 epidemic in France and present the results of three SIR simulations.

Table 6.13 First wave of the COVID-19 epidemic in Spain

Number of prediction/ optimal values of parameters, characteristics	Prediction 1 $n = 14$ Period taken for calculations T_c: March 28–April 10, 2020	Prediction 2 $n = 21$ Period taken for calculations T_c: March 29–April 18, 2020	Prediction 3 $n = 21$ Period taken for calculations T_c: March 21–April 10, 2020
N	736,000	760,152.448	846,400
v	636,763.648	643,334.866338	753,918.9504
α	1.6891621445e-06	1.06816546e-06	2.1002309431e-06
t_1^*	−22.8563373560822	−43.8057685	−13.66667768349
ρ	1.07559704919762	0.6871880836	1.58340390824
$1/\rho$	0.929716198780932	1.455205676	0.6315507968569
r	0.999378854391215	0.997926053	0.9991898592234
F, Eq. (2.9)	9650.57175588669	4566.3928836	11,712.108966599
$F/F_C(1, n-2)$	518.847943864876	300.420584	770.53348464470
S_∞, Eq. (4.9)	546,870	539,142	668,426
V_∞, Eq. (4.16)	189,130	221,011	177,974
t_{final}, Eq. (4.17)	105.6	135.3	93
Final day of the first wave	June 7, 2020	July 7, 2020	May 25, 2020
V_f, taken from [1]	241,550	252,130	235,400
ε, Eq. (4.20)	21.7%	12.3%	24.4%

Optimal values of parameters and other SIR model characteristics for different predictions

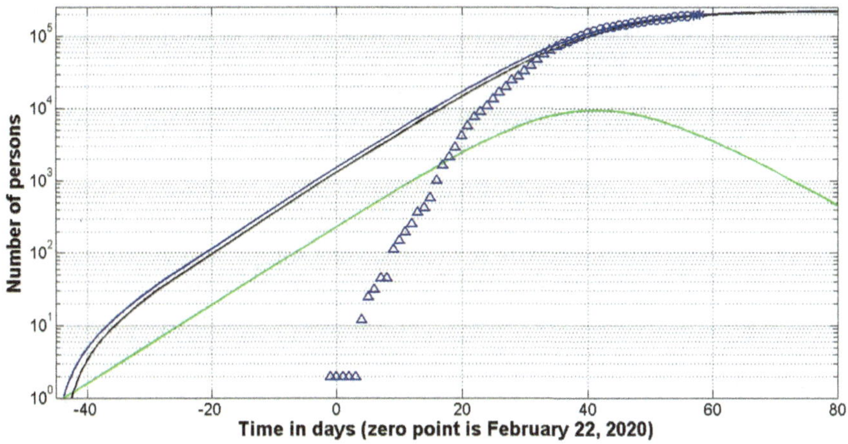

Fig. 6.17 Spain, first epidemic wave, prediction 2: SIR curves (lines) and accumulated number of cases (markers) versus time. Numbers of infected I (green), removed R (black) and the number of victims $V = I + R$ (blue line); "circles" show the cases taken for calculations; "triangles" correspond to the cases during initial stage of the epidemic; "stars"—the last data points used only for a verification of the prediction

SIR simulations for the COVID-19 epidemic dynamics in France were performed with the use of V_j and t_j data sets presented in Tables 6.5 and 6.6 for different time periods T_c taken for calculations. The results are shown in Table 6.14 and Fig. 6.18; some of them were published in [74, 83]. Similar to the case of Spain, the estimations 1–3 for parameters N, ν, α, t_1^* and for the final day of the first wave of the epidemic in France are rather different. The corresponding values of $F/F_C(1, n-2)$ ratio are lower, and the errors ε are higher compared with Spain (see Tables 6.13 and 6.14). Probably, the final size and the duration of the epidemic in France have to be re-estimated with the use of fresher data sets in order to decrease the value of relative error ε.

The smallest value of ε for France corresponds to the prediction 2. Nevertheless, the highest values of $F/F_C(1, n-2)$ and the correlation coefficient r were obtained in prediction 3. Predictions 1 and 3 yield smaller values of the saturation level V_∞ and corresponding V curves deviated faster from the V_j values registered in France after April 10, 2020. Therefore, prediction 2 can be treated as the most reliable and demonstrate once more that more fresh data sets yield better estimations of the COVID-19 epidemic dynamics. The average time of spreading infection $\tau = 1/\rho$ was estimated as 0.81 for this prediction. This value is slightly higher than in Austria (0.77, prediction 6) but much lower than in Italy (1.47, prediction 6) and Spain (1.46, prediction 2). These results can explain much higher numbers of cases

Table 6.14 First wave of the COVID-19 epidemic in France

Number of prediction/optimal values of parameters, characteristics	Prediction 1 $n = 14$ Period taken for calculations T_c: March 28–April 10, 2020	Prediction 2 $n = 14$ Period taken for calculations T_c: April 5–18, 2020	Prediction 3 $n = 21$ Period taken for calculations T_c: March 21–April 10, 2020
N	548,182.4	688,720	449,520
ν	488,472.457423422	621,554.6679951	392,106.816522240
α	2.601817507993e-06	1.970170995e-06	2.98966614701551e-06
t_1^*	−21.1949054853200	−29.7952392	−16.6934804605293
ρ	1.27091619189685	1.2245689787	1.17226847537056
$1/\rho$	0.786833944185958	0.81661385959	0.853046909483677
r	0.996770038705090	0.997453623886	0.998820595668543
F, Eq. (2.9)	1848.61225180740	2347.293599	8040.66641176634
$F/F_C(1, n-2)$	99.3877554735161	126.1985806	528.991211300417
S_∞, Eq. (4.9)	433,257	558,897	339,798
V_∞, Eq. (4.16)	114,925	129,823	109,722
t_{final}, Eq. (4.17)	106.2	119.1	101
Final day of the first wave	June 7, 2020	June 20, 2020	June 2, 2020
V_f, taken from [1]	150,315	154,562	147,935
ε, Eq. (4.20)	23.5%	16.0%	25.8%

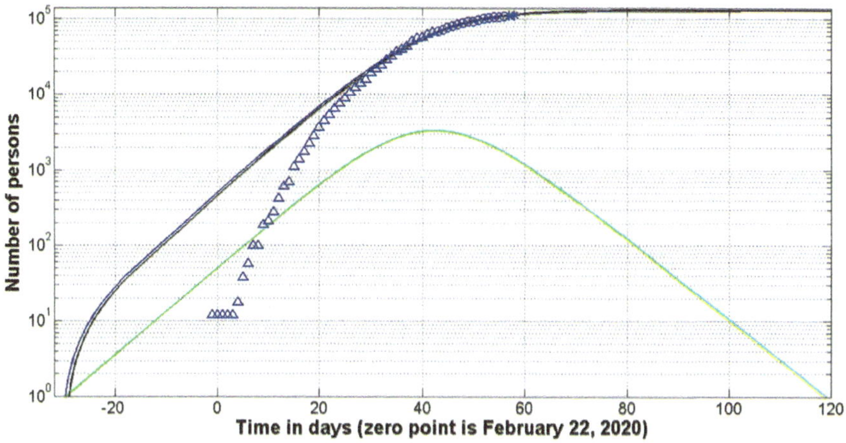

Fig. 6.18 France, first epidemic wave, prediction 2: SIR curves (lines) and accumulated number of cases (markers) versus time. Numbers of infected I (green), removed R (black) and the number of victims $V = I + R$ (blue line); "circles" show the cases taken for calculations; "triangles" correspond to the cases during initial stage of the epidemic; "stars"—the last data points used only for a verification of the prediction

in Spain and Italy during the first epidemic waves. The SIR curves corresponding to prediction 2 are shown in Fig. 6.18.

Figure 6.18 illustrates once more that the registered number of cases during the initial stage of the epidemic outbreak ("triangles") does not reflect the real situation due to many undetected cases. Some estimation of the real number of cases can be the SIR curve $V(t)$ shown in Fig. 6.18. Table 6.14 and Fig. 6.18 show that the epidemic outbreak in France could happen around January 25, 2020. This estimation correlates with the results of paper [85], where the first COVID-19 cases with 5 Chinese tourists are described. They were from Wuhan and arrived in Europe on January 17–22, 2020.

6.7 Germany

In this section, we consider the first wave of the COVID-19 epidemic in Germany and present the results of two SIR simulations.

SIR simulations for the COVID-19 epidemic dynamics in Germany were performed with the use of V_j and t_j data sets presented in Tables 6.3, 6.5 and 6.6 for two different time periods T_c taken for calculations. The results are shown in Table 6.15 and Figs. 6.6, 6.7 and 6.19; some of them were published in [74, 75, 77, 83]. Table 6.15 demonstrates that predictions 1 and 2 yield rather different values of parameters N, v, α, t_1^*, $\tau = 1/\rho$ and for the final day of the first epidemic wave in Germany, but corresponding values of ε are close and rather high (in comparison with Austria, for example). Probably, the final size and the duration of the epidemic in Germany have to be re-estimated taking into account possible new waves of the epidemic, which can occur after the second T_c.

The smallest value of ε, the highest values of $F/F_C(1, n-2)$ and the correlation coefficient r were obtained in prediction 2. Prediction 1 yields smaller values of the saturation level V_∞ and corresponding V curves deviated faster from the V_j values registered in Germany after April 10, 2020. Therefore, prediction 2 can be treated as the most reliable and demonstrate once more that more fresh data sets yield better estimations of the COVID-19 epidemic dynamics. The average time of spreading infection $\tau = 1/\rho$ was estimated as 1.35 for this prediction. This value is much higher than in Austria (0.77, prediction 4) and France (0.82, prediction 2), but slightly lower than in Italy (1.47, prediction 6) and Spain (1.46, prediction 2). The SIR curves corresponding to prediction 1 are shown in Fig. 6.19.

The results for Germany were compared with the situation in other countries and regions shown in Figs. 6.6 and 6.7. The SIR curves (second prediction) and markers representing the V_j values taken for calculations ("circles"), comparisons ("triangles") and verifications of calculations ("stars") are shown in Figs. 6.6 and 6.7 by black color for Germany. It can be seen that the first cases in Germany probably occurred in the beginning of December, 2019 (later than in USA and Italy, but before South Korea and the UK). In May 2020, the numbers of cases in Germany, Spain, Italy, and the UK were rather close.

Table 6.15 First wave of the COVID-19 epidemic in Germany

Number of prediction/optimal values of parameters, characteristics	Prediction 1 $n = 14$ Period taken for calculations T_c: March 28–April 10, 2020	Prediction 2 $n = 21$ Period taken for calculations T_c: April 9–29, 2020
N	1,023,648	882,400
ν	946,949.0	790,630.4
α	1.9664814e-06	9.3698628e-07
t_1^*	−22.83551	−74.7002
ρ	1.862158	0.7408098
$1/\rho$	0.537011	1.349874
r	0.9990431	0.999140468
F, Eq. (2.9)	6261.40324	11,038.27866
$F/F_C(1, n-2)$	336.63458	726.20254
S_∞, Eq. (4.9)	874,169	705,448
V_∞, Eq. (4.16)	149,479	176,952
t_{final}, Eq. (4.17)	104.8	164.2
Final day of the first wave	June 6, 2020	August 4, 2020
V_f, taken from [1]	183,979	212,022
ε, Eq. (4.20)	18.8%	16.5%

Optimal values of parameters and other SIR model characteristics for different predictions

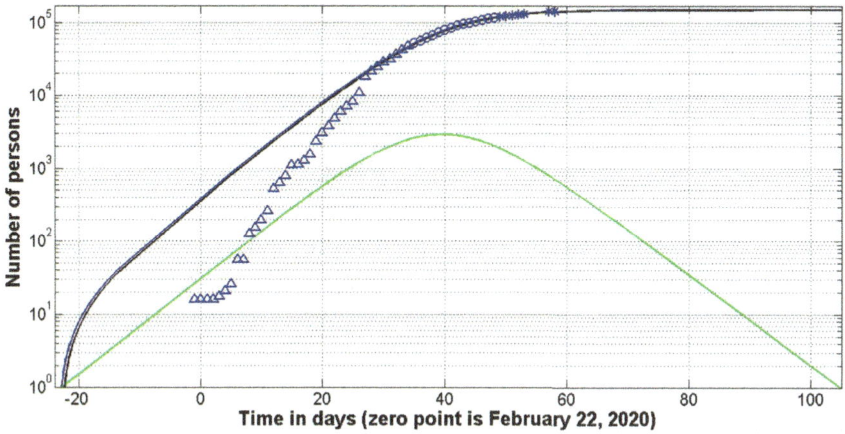

Fig. 6.19 Germany, first epidemic wave, prediction 1: SIR curves (lines) and accumulated number of cases (markers) versus time. Numbers of infected I (green), removed R (black) and the number of victims $V = I + R$ (blue line); "circles" show the cases taken for calculations; "triangles" correspond to the cases during initial stage of the epidemic; "stars"—the last data points used only for a verification of the prediction

6.8 The UK

In this section, we consider the first wave of the COVID-19 epidemic in the UK and present the results of a SIR simulation.

The SIR simulation for the COVID-19 epidemic dynamics in the UK was performed with the use of V_j and t_j data sets presented in Table 6.3 for period of time T_c form April 9 to April 29, 2020, taken for calculations. The results are shown in Table 6.16 and Figs. 6.6, 6.7, 6.20 and 6.21; some of them were published in [75, 77]. The value V_f was interpolated with the use of figures for accumulated number of cases in the UK—324,605 (as of August 22, 2020) and weekly increase 7161 available in WHO report [82]. Unfortunately, WHO stopped to present daily information after August 16, 2020.

Table 6.16 demonstrates high enough values of $F/F_C(1, n-2)$ and the correlation coefficient r (comparable with the presented estimations for the EU countries), but the relative error is much higher ($\varepsilon = 33.4\%$). As quarantine restrictions in the UK were introduced later than in EU countries, many cases of the disease could be undetected in the UK in April 2020. The average time of spreading infection $\tau = 1/\rho$ in UK was estimated as 3.03 days. This value is much higher than in the EU countries: Austria (0.77, prediction 4), France (0.82, prediction 2), Germany (1.35, prediction 2), Italy (1.47, prediction 6), and Spain (1.46, prediction 2). This may also be the result of the later introduction of quarantine in the UK.

Figures 6.20 and 6.21 illustrate that the number of cases registered in the UK after T_c ("stars") started to deviate from the theoretical $V = I + R$ curve (solid line) immediately after T_c. The result of this behavior is the rather large value of the

Table 6.16 First wave of the COVID-19 epidemic in the UK

Number of prediction/optimal values of parameters, characteristics	Prediction 1 $n = 21$ Period taken for calculations T_c: April 9–29, 2020
N	479,782.4
v	361,758.017613005
α	9.1371956639e-07
t_1^*	−47.0679605620665
ρ	0.330545378991741
$1/\rho$	3.02530322175518
r	0.999363727388945
F, Eq. (2.9)	14,916.4586251212
$F/F_C(1, n-2)$	981.345962179029
S_∞, Eq. (4.9)	264,883
V_∞, Eq. (4.16)	214,899
t_{final}, Eq. (4.17)	180.1
Final day of the first wave	August 20, 2020
V_f, estimated from [86]	322,559
ε, Eq. (4.20)	33.4%

Optimal values of parameters and other SIR model characteristics

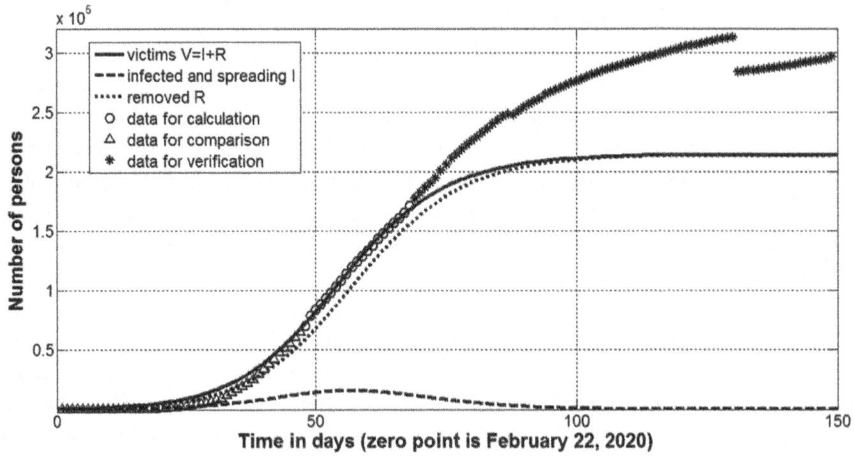

Fig. 6.20 The UK, first epidemic wave, prediction 1: SIR curves (lines) and accumulated number of cases (markers) versus time (linear scale). Numbers of infected I (dashed), removed R (dotted) and the number of victims $V = I + R$ (solid line); "circles" show the cases taken for calculations; "triangles" correspond to the cases during initial stage of the epidemic; "stars"—the last data points used only for a verification of the prediction [1]

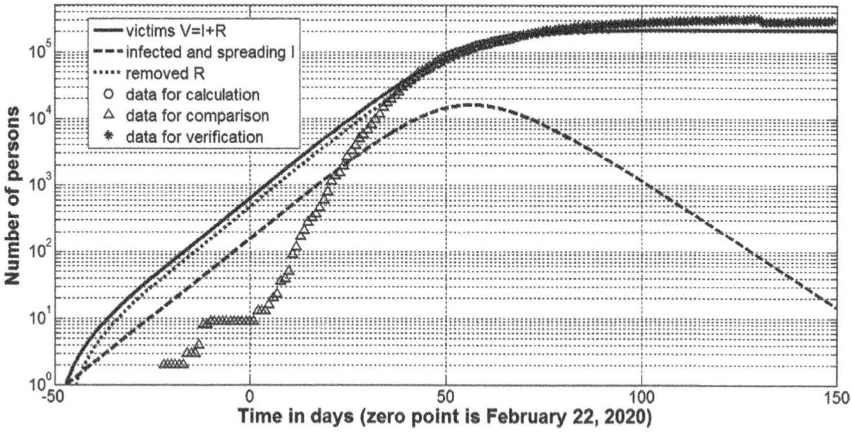

Fig. 6.21 The UK, first epidemic wave, prediction 1: SIR curves (lines) and accumulated number of cases (markers) versus time (logarithmic scale). Numbers of infected I (dashed), removed R (dotted) and the number of victims $V = I + R$ (solid line); "circles" show the cases taken for calculations; "triangles" correspond to the cases during initial stage of the epidemic; "stars"—the last data points used only for a verification of the prediction [1]

relative error ε. Usually, the accuracy of predictions can be improved with the use of fresher data sets, but probably it does not work in the case of the UK statistics due to its very irregular character. In particular, sometimes the official accumulated number of cases (reported in [1]) decreases by a jump (see Fig. 6.20).

A comparison of the epidemic dynamics in the UK with the pandemic situation in other countries and in the world is shown in Figs. 6.6 and 6.7. The SIR curves and markers representing the V_j values taken for calculations ("circles"), comparisons ("triangles") and verifications of calculations ("stars") are shown by red color for the UK. It can be seen that the first cases in the UK probably occurred in the beginning of January, 2020 (later than in USA, Italy, Germany and South Korea, and approximately in the same time with Spain). But on May 13, 2020, the number of cases registered in the UK was higher than in the EU countries (according to [1]).

6.9 USA

In this section, we consider the first wave of the COVID-19 epidemic in USA and present the results of a SIR simulation.

The SIR simulation for the COVID-19 epidemic dynamics in USA was performed with the use of V_j and t_j data sets presented in Table 6.3 for period of time T_c from April 9 to April 29, 2020, taken for calculations. The results are shown in Table 6.17 and Figs. 6.6, 6.7, 6.22 and 6.23; some of them were published in [75, 77]. The value V_f was interpolated using figures for accumulated number of cases in USA 6,960,152 (on September 26, 2020) and weekly increase 298,149 available in WHO report [86].

Table 6.17 demonstrates high enough values of $F/F_C(1, n-2)$ and the correlation coefficient r (comparable with the presented estimations for the EU countries), but the relative error is much higher (ε = 78.2%). As quarantine restrictions in USA were introduced later than in EU countries (see, e.g., [87]), many cases of the disease could be undetected in USA in April 2020. The error could be reduced with the use of fresher data sets. The time period between T_c and t_{final} is very long (almost 5 months) in the case of USA. During this period, the characteristics of the epidemic probably changed (due to changes in quarantine conditions and mass protests in the summer of 2020), so the applied classical SIR theory (see Chap. 4) and method of determining its parameters (proposed in Chap. 5) are not suitable for adequate simulation of the situation in the USA. Probably, new COVID-19 epidemic waves started in this country before the first wave ended.

The great value of the error ε calls into question the obtained estimates of the characteristics of the epidemic. The average time of spreading infection $\tau = 1/\rho$ in USA was estimated as 1.45 days. This value is higher than in Austria (0.77, prediction 6), France (0.82, prediction 2), and Germany (1.35, prediction 2), comparable with Italy (1.47, prediction 6) and Spain (1.46, prediction 2) and lower than in the UK (3.03).

Table 6.17 First wave of the COVID-19 epidemic in USA

Number of prediction/optimal values of parameters, characteristics	Prediction 1 $n = 21$ Period taken for calculations T_c: April 9–29, 2020
N	7,082,400
v	6,312,894.63367680
α	1.08933738707e-07
t_1^*	−85.9627117094321
ρ	0.687687214511608
$1/\rho$	1.45414947216984
r	0.999378778826037
F, Eq. (2.9)	15,278.2115499352
$F/F_C(1, n-2)$	1005.14549670626
S_∞, Eq. (4.9)	5,601,247
V_∞, Eq. (4.16)	1,481,153
t_{final}, Eq. (4.17)	213.5
Final day of the first wave	September 22, 2020
V_f, estimated from [86]	6,789,781
ε, Eq. (4.20)	78.2%

Optimal values of parameters and other SIR model characteristics

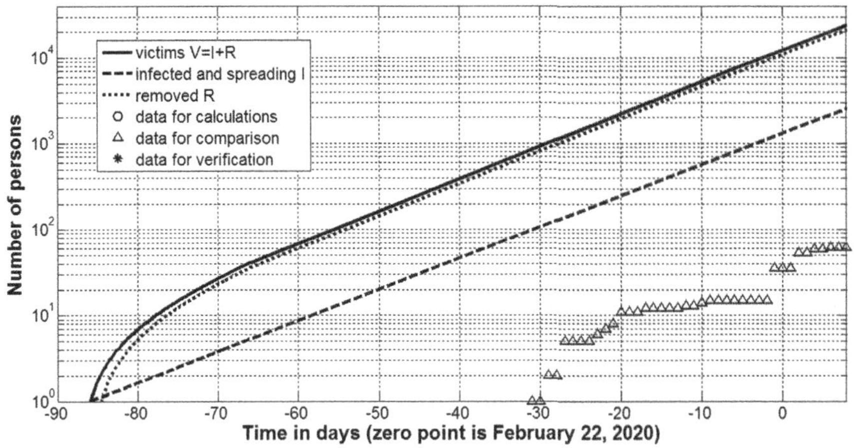

Fig. 6.22 USA, first epidemic wave, prediction 1: SIR curves (lines) and accumulated number of cases (markers) versus time before March 1, 2020. Numbers of infected I (dashed), removed R (dotted) and the number of victims $V = I + R$ (solid line); "triangles" correspond to the registered number of cases during the initial stage of the epidemic

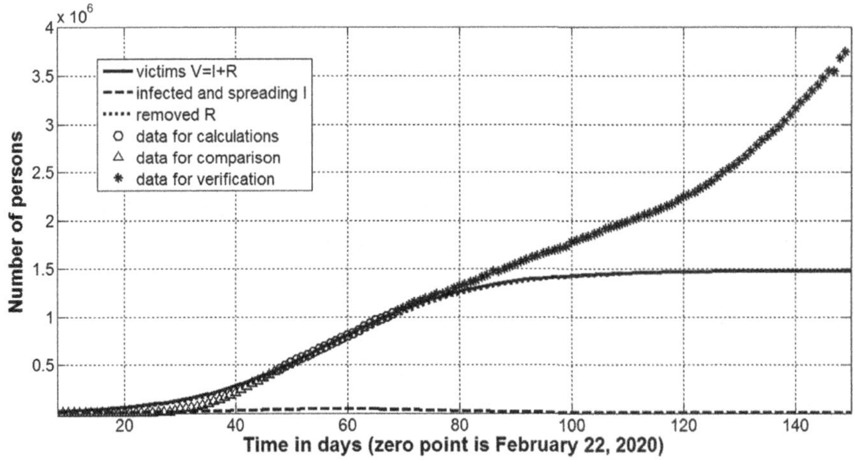

Fig. 6.23 USA, first epidemic wave, prediction 1: SIR curves (lines) and accumulated number of cases (markers) versus time after March 1, 2020. Numbers of infected I (dashed), removed R (dotted) and the number of victims $V = I + R$ (solid line); "circles" show the cases taken for calculations; "triangles" correspond to the cases during initial stage of the epidemic; "stars"—the last data points used only for a verification of the prediction

The basic reproduction number R_0 can be calculated with the use of formula (4.33) and the estimation of the doubling time $\tau_d = 2.31$ days obtained in Chap. 2 using the assumption that the number of cases increased exponentially. With the use of the estimation $\tau = 1/\rho = 1.454$ days (see Table 6.17), Eq. (4.33) yields

$R_0 = 1.55$. The basic reproduction numbers of the COVID-19 pandemic were estimated in [12–14, 22, 30]. According to WHO calculations for China [12], values R_0 vary from 2 to 2.5.

Figure 6.22 illustrate that the number of cases registered in USA before March 1, 2020 (V_j, "triangles") was significantly less than the theoretical estimate $V = I + R$ (solid line). For almost two months, the COVID-19 epidemic remained unnoticed in this country. This conclusion is in good agreement with the results of a very interesting study [88]. A fundamental statistical analysis showed that the number of cases of pneumonia in the USA in January and February 2020 exceeded previous year's figures and this excess was higher in those states where the actual numbers of reported COVID-19 cases are larger [88].

Figure 6.23 illustrates that the number of cases registered in USA (V_j, "stars") started to deviate significantly from the theoretical $V = I + R$ curve (solid line) in 10–15 days after T_c. The result of this behavior is the very large value of the relative error ε. Usually, the accuracy of predictions can be improved with the use of fresher data sets, but the character of the increase in the number of cases V_j demonstrates that next waves of the epidemic occurred. It means that classical SIR theory (presented in Chap. 4) and the identification of its parameters (proposed in Chap. 5 and used in this chapter to calculate SIR curves) does not work in the case of USA statistics. Later, we will discuss how the new waves of the pandemic can be simulated with the use of generalized SIR model.

A comparison of the epidemic dynamics in USA with the pandemic situation in other countries and in the world is shown in Figs. 6.6 and 6.7. The SIR curves and markers representing the V_j values taken for calculations ("circles"), comparisons ("triangles") and verifications of calculations ("stars") are shown by blue color for USA. Figure 6.6 demonstrates that the first COVID-19 cases in the USA probably occurred at the same time as in Italy (compare blue and green lines). Before March 1, 2020, the estimated numbers of cases in Italy and USA were rather close. Nevertheless, on March 1 the number of confirmed cases in Italy was 1689 in comparison with 62 in USA. This delay with detection of infected persons and later quarantining (see, e.g., [87]; in particular, in New York the quarantine was announced on March 20, in comparison with March 8 in Lombardy and March 10 in Italy) caused the huge difference in number of cases (visible in Fig. 6.7). Figure 6.6 shows that in the Republic of Korea the detected number of cases was close to the estimated one (compare solid magenta line and magenta "triangles") already on March 1. Timely testing and isolation of patients allowed this country to avoid a large number of cases (visible in Fig. 6.7).

6.10 The Republic of Moldova

In this section, we consider the first wave of the COVID-19 epidemic in the Republic of Moldova and present the results of two SIR simulations.

SIR simulations for the COVID-19 epidemic dynamics in Moldova were performed with the use of V_j and t_j data sets presented in Tables 6.5 and 6.6 for two different time periods T_c taken for calculations. The results are shown in Table 6.18 and Fig. 6.24; some of them were published in [74, 83]. Table 6.18 demonstrates that both predictions were too optimistic. The real number of cases in this country exceeded the predicted saturation levels V_∞ which are rather different for predictions 1 and 2.

Despite rather high values of $F/F_C(1, n-2)$ and the correlation coefficient r, the relative errors are very high ($\varepsilon = 71$–77%). These figures are similar to the case of USA ($\varepsilon = 78.2\%$), but the accuracy of predictions for the EU countries (calculated with the same time periods T_c) is much better. The reason may be problems with testing, detection and isolation of patients in the Republic of Moldova in March–April 2020. These difficulties occurred everywhere at the beginning of the epidemic, but comparing the accuracy of SIR simulations for Moldova and the EU countries indicates better organization to combat the pandemic in European Union.

The great value of the error ε calls into question the obtained estimates of the characteristics of the epidemic. The average times of spreading infection $\tau = 1/\rho$ in Moldova were estimated as 5.45 and 3.23 days for the first and second prediction, respectively. These values are higher than in the EU countries (0.77–1.47), USA (1.45), and the UK (3.03). It can be stated that SIR simulation allows us to assess

Table 6.18 First wave of the COVID-19 epidemic in the Republic of Moldova

Number of prediction/optimal values of parameters, characteristics	Prediction 1 $n = 14$ Period taken for calculations T_c: March 28–April 10, 2020	Prediction 2 $n = 14$ Period taken for calculations T_c: April 5–18, 2020
N	2911.36	6079.36
v	1322.03320401920	4055.26811705
α	0.0001386860412	7.641517427e-05
t_1^*	12.9079807669692	4.3218081122
ρ	0.18334755141242	0.30988401
$1/\rho$	5.45412247012025	3.2270137723
r	0.99726948597323	0.998814863
F, Eq. (2.9)	2188.39255686532	5053.707894
$F/F_C(1, n-2)$	117.655513809963	271.7047254
S_∞, Eq. (4.9)	453	2539
V_∞, Eq. (4.16)	2458	3541
t_{final}, Eq. (4.17)	109.9	114.5
Final day of the first wave	June 11, 2020	June 16, 2020
V_f, taken from [1]	10,727	12,254
ε, Eq. (4.20)	77.1%	71.1%

Optimal values of parameters and other SIR model characteristics for different predictions

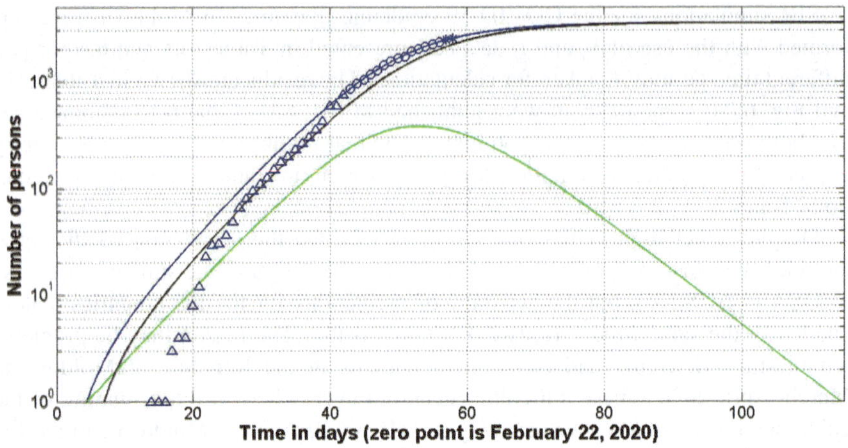

Fig. 6.24 Moldova, first epidemic wave, prediction 2: SIR curves (lines) and accumulated number of cases (markers) versus time. Numbers of infected I (green), removed R (black) and the number of victims $V = I + R$ (blue line); "circles" show the cases taken for calculations; "triangles" correspond to the cases during initial stage of the epidemic; "stars"—the last data points used only for a verification of the prediction

the state of detection and isolation of patients in real time and make timely suggestions for improving the situation.

Figure 6.24 shows the results of calculations with the use of the values of optimal parameters for prediction 2. It can bee seen that the hidden period in Moldova (distance between t_1^* and the day when the first COVID-19 case was laboratory confirmed) was estimated to be not very long. But the low accuracy of the predictions casts doubt on this conclusion. It is likely that new forecasts using more recent data will give a longer duration of the hidden period for Moldova.

6.11 Ukraine

In this section, we consider the first wave of the COVID-19 epidemic in Ukraine and present the results of several SIR simulations.

SIR simulations for the COVID-19 epidemic dynamics in Ukraine were performed with the use of V_j and t_j data sets presented in Tables 6.11, 6.19, 6.20 and 6.22 for different time periods T_c taken for calculations. Some results were published in [71–73, 76–78].

The first laboratory-confirmed COVID-19 case in Ukraine was registered on March 3, 2020 (see Table 6.19). The first SIR simulations have been performed with the use of data sets corresponding to the epidemic beginning. The results are presented in Table 6.21 and Figs. 6.25 and 6.26. All three predictions presented in Table 6.21 were very optimistic, but the actual number of cases V_j recorded in

Table 6.19 Official cumulative numbers of confirmed COVID-19 cases in Ukraine after the epidemic outbreak [1]

Day in March 2020	Time moments in days t_j	Number of cases V_j	Day in March and April 2020	Time moments in days t_j	Number of cases V_j
3	0	1	24	21	113
4	1	1	25	22	156
5	2	1	26	23	218
6	3	1	27	24	311
7	4	1	28	25	418
8	5	1	29	26	480
9	6	1	30	27	549
10	7	1	31	28	669
11	8	1	1	29	804
12	9	3	2	30	987
13	10	3	3	31	1096
14	11	3	4	32	1251
15	12	3	5	33	1319
16	13	7	6	34	1462
17	14	14	7	35	1668
18	15	16	8	36	1892
19	16	16	9	37	2203
20	17	26	10	38	2511
21	18	47	11	39	2777
22	19	47	12	40	3102
23	20	84	–	–	–

Ukraine after the period T_c very quickly exceeded the corresponding theoretical predictions (see, e.g., Fig. 6.25). As a result, the accuracy was very low (ε = 69–77%, similar to the cases of Moldova and USA), despite rather high values of $F/F_C(1, n-2)$ and the correlation coefficient r.

The reason should be sought in poor organizations of testing, detection and isolation of patients in March–April 2020 and may be later as well. For example, in Ukraine the PLC tests were introduced for pneumonia patients and the medical staff only after April 10, 2020. Predictions 2 and 3 could help to estimate how the change in testing algorithm affected the epidemic dynamics in Ukraine, since they were performed with the use of data sets for the periods March 28–April 10 and April 11–24. Delayed proper testing could cost Ukraine at least 9000 additional patients and an epidemic duration increase of 47 days (compare results of predictions 2 and 3 in Table 6.21). That is why the maximum PCR testing (especially for medical staff and patients with pneumonia) can be recommended in all countries as an effective means of reducing the scale and duration of the pandemic.

Table 6.20 Official cumulative numbers of confirmed COVID-19 cases in Ukraine and Kyiv [1, 73, 80]

Day in March and April 2020	Time moments in days t_j	Number of cases in Kyiv	Day in April 2020	Time moments in days t_j	Number of cases in Kyiv	Number of cases in Ukraine
19	16	3	11	39	416	2777
20	17	10	12	40	495	3102
21	18	10	13	41	551	3372
22	19	28	14	42	612	3764
23	20	31	15	43	644	4162
24	21	34	16	44	705	4662
25	22	40	17	45	770	5106
26	23	55	18	46	801	5449
27	24	55	19	47	850	5710
28	25	82	20	48	892	6125
29	26	102	21	49	989	6592
30	27	107	22	50	1059	7170
31	28	132	23	51	1122	7647
1	29	161	24	52	1159	8125
2	30	189	25	53	1188	8617
3	31	214	26	54	1220	9009
4	32	245	27	55	1298	9410
5	33	267	28	56	1341	9866
6	34	279	29	57	1401	10,406
7	35	294	–	–	–	–
8	36	335	–	–	–	–
9	37	348	–	–	–	–
10	38	396	–	–	–	–

Figures 6.25 and 6.26 illustrate the results of estimations 2 and 3. Figure 6.25 shows that even prediction 3 was too optimistic. New cases registered after corresponding period T_c and shown by "stars" deviate from the solid blue line very fast. Unfortunately, new calculations (performed with the use of fresher data sets (e.g., shown in Table 6.22)) yielded similar results. The results of some of these forecasts are shown in the Tables 5.2 and 6.23 and Figs. 6.15 and 6.16. Figure 6.16 shows how quickly the results of observations V_j (blue "stars") deviate from the theoretical dependence $V(t)$ (blue solid line). The obtained results and comparisons with the above forecasts for the EU countries allow us to conclude that the premature easing of quarantine in Ukraine led to the beginning of the second and subsequent waves of the epidemic before the end of the first wave. Probably, we have similar dynamics in the UK, Moldova, USA and the world. The next waves of the pandemic will be discussed in the following chapters.

Table 6.21 First wave of the COVID-19 epidemic in Ukraine. Predictions 1–3

Number of prediction/optimal values of parameters, characteristics	Prediction 1 $n = 14$ Period taken for calculations T_c: March 13–26, 2020	Prediction 2 $n = 14$ Period taken for calculations T_c: March 28–April 10, 2020	Prediction 3 $n = 14$ Period taken for calculations T_c: April 11–24, 2020
N	8928.736	10,463.808	21,380.92416
v	8020.1295	6672.48742023168	12,190.2413389748
α	0.0003422	4.50100779399318e-05	1.45792603472439e-05
t_1^*	12.578686	5.04959595059797	-5.72672361552338
ρ	2.7446302	0.300329178837842	0.177724702176649
$1/\rho$	0.3643478	3.32967979957729	5.62667984671064
r	0.99827735	0.996730333110634	0.998863416866565
F, Eq. (2.9)	3474.00881	1826.05423193279	5269.98203160936
$F/F_C(1, n-2)$	186.774667	98.1749587060637	283.332367290826
S_∞, Eq. (4.9)	7168	3930	6108
V_∞, Eq. (4.16)	1761	6534	15,273
t_{final}, Eq. (4.17)	47.4	97	145
Final day of the first wave	April 19, 2020	June 8, 2020	July 26, 2020
V_f, taken from [1]	5710	27,856	65,656
ε, Eq. (4.20)	69.2	76.5%	76.7%

Optimal values of parameters and other SIR model characteristics for different predictions

Great values of the error ε call into question the obtained estimates of the characteristics of the epidemic. The average times of spreading infection $\tau = 1/\rho$ in Ukraine vary from 1 to 21 days (see Tables 5.2, 6.21 and 6.23). For example, the figure $\tau = 5$ (corresponding to prediction 8) is much higher than in the EU countries (0.77–1.47), USA (1.45), and the UK (3.03) and comparable with the results for Moldova (3.2 and 5.5).

6.12 Kyiv

In this section, we consider the first wave of the COVID-19 epidemic in the city of Kyiv, present and discuss the results of some SIR simulations.

SIR simulations for the COVID-19 epidemic dynamics in the capital of Ukraine were performed with the use of V_j and t_j data sets presented in Table 6.20 for different time periods T_c taken for calculations. Some results were published in [73, 76, 77]. We planned to conduct similar studies for other Ukrainian regions [90], but our application to the National Research Foundation of Ukraine was rejected.

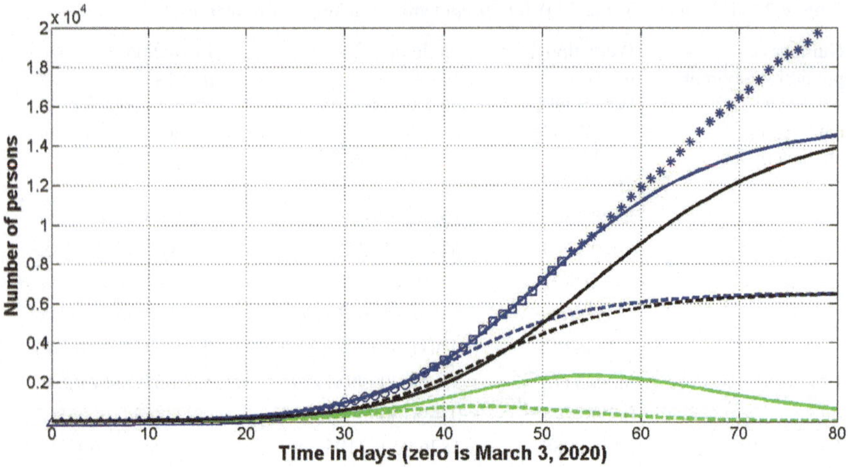

Fig. 6.25 Ukraine, first epidemic wave, predictions 2 and 3 (linear scale): SIR curves (lines) and accumulated number of cases (markers) versus time. Numbers of infected and spreading I (green), removed R (black) and victims $V = I + R$ (blue). Dashed lines correspond to the prediction 2 (T_c: March 28–April 10; corresponding values are shown by "circles"), solid—to the prediction 3 (T_c: April 11–24; corresponding values are shown by "squares"). "Triangles" show the accumulated numbers of cases before March 28, "stars"—after April 10

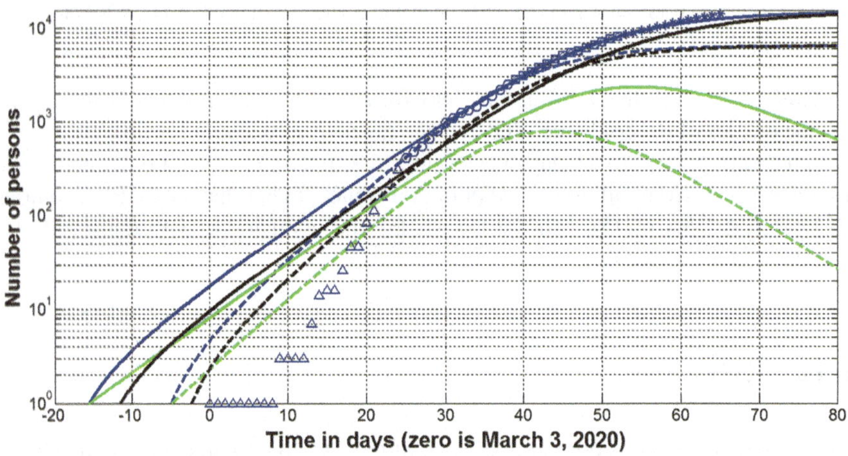

Fig. 6.26 Ukraine, first epidemic wave, predictions 2 and 3 (logarithmic scale): SIR curves (lines) and accumulated number of cases (markers) versus time. Numbers of infected and spreading I (green), removed R (black) and victims $V = I + R$ (blue). Dashed lines correspond to the prediction 2 (T_c: March 28–April 10; corresponding values are shown by "circles"), solid—to the prediction 3 (T_c: April 11–24; corresponding values are shown by "squares"). "Triangles" show the accumulated numbers of cases before March 28, "stars"—after April 10

Table 6.22 Official cumulative numbers of confirmed COVID-19 cases in Ukraine and in the world [1, 89]

Day in April, May and June 2020	Time in days t_j	Number of cases in Ukraine	Global number of cases	Day in June and July 2020	Time in days t_j	Number of cases in Ukraine	Global number of cases
29	100	10,406	3,175,207	14	146	31,810	7,823,289
30	101	10,861	3,267,184	15	147	32,476	7,941,791
1	102	11,411	3,349,786	16	148	33,234	8,061,550
2	103	11,913	3,435,894	17	149	34,063	8,242,999
3	104	12,331	3,517,345	18	150	34,984	8,385,440
4	105	12,697	3,588,773	19	151	35,825	8,525,042
5	106	13,184	3,672,238	20	152	36,560	8,708,008
6	107	13,691	3,759,967	21	153	37,241	8,860,331
7	108	14,195	3,855,788	22	154	38,074	8,993,659
8	109	14,710	3,917,366	23	155	39,014	9,129,146
9	110	15,232	4,006,257	24	156	40,009	9,296,202
10	111	15,648	4,088,848	25	157	41,117	9,473,214
11	112	16,023	4,170,424	26	158	42,065	9,653,048
12	113	16,425	4,248,389	27	159	42,982	9,843,073
13	114	16,847	4,338,658	28	160	43,628	10,021,401
14	115	17,330	4,425,485	29	161	44,334	10,185,374
15	116	17,858	4,525,497	30	162	44,998	10,357,662
16	117	18,291	4,618,821	1	163	45,888	10,533,779
17	118	18,616	4,731,458	2	164	46,763	10,710,005
18	119	18,876	4,789,205	3	165	47,677	10,922,324
19	120	19,230	4,893,186	4	166	48,500	11,125,245
20	121	19,706	4,993,470	5	167	49,043	11,327,790
21	122	20,148	5,103,006	6	168	49,607	11,500,302
22	123	20,580	5,204,508	7	169	50,414	11,669,259
23	124	20,986	5,304,772	8	170	51,224	11,874,226
24	125	21,245	5,404,512	9	171	52,043	12,102,328
25	126	21,584	5,488,825	10	172	52,843	12,322,395
26	127	21,905	5,593,631	11	173	53,521	12,552,765
27	128	22,382	5,701,337	12	174	54,133	12,768,307
28	129	22,811	5,817,385	13	175	54,771	12,964,809
29	130	23,294	5,934,936	14	176	55,607	13,150,645
30	131	23,672	6,057,853	15	177	56,455	13,378,853
31	132	24,002	6,194,533	16	178	57,264	13,616,593
1	133	24,340	6,287,771	17	179	58,111	13,876,441
2	134	24,823	6,416,828	18	180	58,842	14,043,176
3	135	25,411	6,535,354	19	181	59,493	14,348,858

(continued)

Table 6.22 (continued)

Day in April, May and June 2020	Time in days t_j	Number of cases in Ukraine	Global number of cases	Day in June and July 2020	Time in days t_j	Number of cases in Ukraine	Global number of cases
4	136	25,964	6,663,304	20	182	60,166	14,562,550
5	137	26,514	6,663,304	21	183	60,995	14,765,256
6	138	26,999	6,799,713	22	184	61,851	15,012,731
7	139	27,462	6,931,000	23	185	62,823	15,296,926
8	140	27,856	7,039,918	24	186	63,929	15,581,009
9	141	28,381	7,145,539	25	187	64,849	15,785,641
10	142	29,070	7,273,958	26	188	65,656	16,114,449
11	143	29,753	7,410,510	27	189	66,575	16,341,920
12	144	30,506	7,553,182	28	190	67,597	16,558,289
13	145	31,154	7,690,708	29	191	68,794	16,812,755

Table 6.23 First wave of the COVID-19 epidemic in Ukraine. Predictions 7 and 8

Number of prediction/optimal values of parameters, characteristics	Prediction 7 $n = 14$ Period taken for calculations T_c: May 13–26, 2020	Prediction 8 $n = 14$ Period taken for calculations T_c: May 3–16, 2020
N	92,770.3987200000	59,002.368
v	76,647.6354922344	43,307.0168670536
α	3.766268949867e-06	4.62451395337e-06
t_1^*	−32.4672286400970	−16.5560950369874
ρ	0.288675609635095	0.200273903780639
$1/\rho$	3.46409591466375	4.99316177056849
r	0.998856483326519	0.999453409358706
F, Eq. (2.9)	5237.97374486281	10,968.1371753815
$F/F_C(1, n-2)$	281.611491659291	589.684794375351
S_∞, Eq. (4.9)	62,507	30,676
V_∞, Eq. (4.16)	30,263	28,327
t_{final}, Eq. (4.17)	273.6	264.9
Final day of the first wave	October 20, 2020	October 11, 2020
V_f	315,826	265,454
ε, Eq. (4.20)	90.4%	89.3%

Optimal values of parameters and other SIR model characteristics for different predictions

Table 6.24 First wave of the COVID-19 epidemic in Kyiv

Number of prediction/optimal values of parameters, characteristics	Prediction 1 $n = 14$ Period taken for calculations T_c: March 28–April 10, 2020	Prediction 2 $n = 14$ Period taken for calculations T_c: April 11–24, 2020	Prediction 3 $n = 14$ Period taken for calculations T_c: May 13–26, 2020
N	3572.256	3025.6064	3916.224
v	3287.84315106509	1887.90132535378	1901.88529953178
α	0.00064332173	0.00011892581484817	6.31210877007e-05
t_1^*	13.4243281820969	−1.03084215697423	6.20323541303859
ρ	2.11514094406513	0.224520203470638	0.120049068788449
$1/\rho$	0.472781732492	4.45394215995700	8.32992717138191
r	0.995627168903	0.996747099239341	0.996899503409035
F, Eq. (2.9)	1363.11536701815	1835.51243347698	1926.17853220408
$F/F_C(1, n-2)$	73.2857724203304	98.6834641654292	126.722271855531
S_∞, Eq. (4.9)	3009	1078	735
V_∞, Eq. (4.16)	563	1947	3181
t_{final}, Eq. (4.17)	55	121	172
Final day of the first wave	April 27, 2020	July 2, 2020	August 22, 2020
V_f, taken from [80]	1298	5347	11,804
ε, Eq. (4.20)	56.6%	63.6%	73.1%

Optimal values of parameters and other SIR model characteristics for different predictions

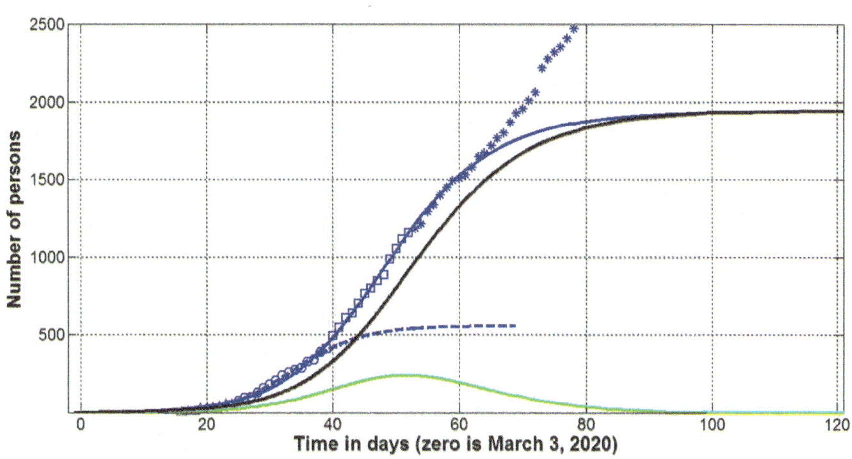

Fig. 6.27 Kyiv, first epidemic wave, predictions 1 and 2: SIR curves (lines) and accumulated number of cases (markers) versus time (linear scale). Numbers of infected and spreading I (green), removed R (black) and victims $V = I + R$ (blue). Dashed line corresponds to prediction 1 (calculated with the use of values V_j for T_c: March 28–April 10, shown by "circles"), solid line—to prediction 2 (calculated with the use of values V_j for T_c: April 11–24, shown by "squares"). "Triangles" represent the accumulated numbers of cases before March 28, "stars"—after April 10

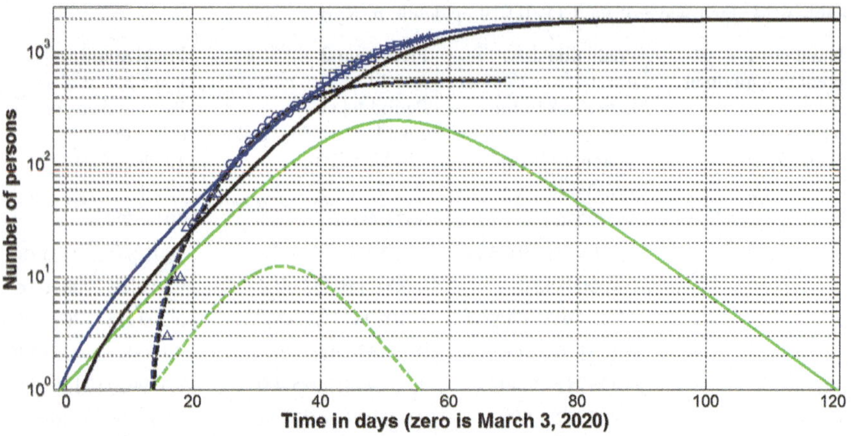

Fig. 6.28 Kyiv, first epidemic wave, predictions 1 and 2: SIR curves (lines) and accumulated number of cases (markers) versus time (logarithmic scale). Numbers of infected and spreading I (green), removed R (black) and victims $V = I + R$ (blue). Dashed lines correspond to prediction 1 (calculated with the use of values V_j for T_c: March 28–April 10, shown by "circles"), solid line—to prediction 2 (calculated with the use of values V_j for T_c: April 11–24, shown by "squares"). "Triangles" represent the accumulated numbers of cases before March 28, "stars"—after April 10

The first laboratory confirmed COVID-19 case in Kyiv was registered on March 19, 2020 (see Tables 6.20 and 5.1). Since in Ukraine the PLC tests were introduced for pneumonia patients and the medical staff only after April 10, 2020, it was interesting to investigate how the new testing algorithm changed the epidemic dynamics in Kyiv. For this purpose, we have used two data sets for the periods March 28–April 10 and April 11–24. The results of SIR simulations are presented in Table 6.24 (predictions 1 and 2) and Figs. 6.27 and 6.28.

Predictions presented in Table 6.24 were very optimistic, but the actual number of cases V_j recorded in Kyiv after periods T_c very quickly exceeded the corresponding theoretical predictions (see, e.g., Figure 6.27). As a result, the accuracy was very low (ε = 57–64%, similar to the case of Ukraine), despite rather high values of $F/F_C(1, n-2)$ and the correlation coefficient r. The reason should be sought in poor organizations of testing, detection and isolation of patients in March–April 2020 and may be later as well. For example, in Ukraine the PLC tests were introduced for pneumonia patients and the medical staff only after April 10, 2020. Predictions 1 and 2 could help to estimate how the change in testing algorithm affected the epidemic dynamics in Kyiv, since they were performed with the use of data sets for the periods March 28–April 10 and April 11–24. Delayed proper testing could cost the city of Kyiv at least 1400 additional patients and an epidemic duration increase of 66 days (compare results of predictions 1 and 2 in Table 6.24).

Such a large number of COVID-19 patients who were not detected and isolated in time in Ukraine and Kyiv led to a large-scale spread of the epidemic. As of November 11, 2020, the number of cases in Ukraine exceeded half a million, and in Kyiv 47 thousand [80].

Figures 6.27 and 6.28 illustrate the results of estimations 1 and 2. Figure 6.27 shows that even prediction 2 was too optimistic. New cases registered after corresponding period T_c (shown by "stars") deviate from the solid blue line very fast. Unfortunately, new calculations (performed with the use of fresher data sets (one of them is shown in Table 6.24, prediction 3)) yielded similar results. In particular, the accuracy of prediction 3 ($\varepsilon = 73.1$) was even worse than for predictions 1 and 2.

The obtained results and comparisons with the above forecasts for Ukraine confirm the conclusion that the quarantine in Ukraine and its capital Kyiv was relaxed too early (in comparison with the EU countries). The epidemic dynamics in Kyiv is similar to ones in Ukraine, the UK, Moldova, USA, and the world. Second and next epidemic waves in Kyiv started before the first one finished.

Great values of the error ε call into question the obtained estimates of the epidemic characteristics. The average times of spreading infection $\tau = 1/\rho$ in Kyiv vary from 0.5 to 8.3 days (see Table 6.24). These figures correspond to ones obtained for Ukraine and Moldova, but probably are much higher than in the EU countries (0.77–1.47), USA (1.45), and the UK (3.03).

The analysis of the COVID-19 epidemic dynamics in Kyiv allows us to discuss the question: Could the pandemic be avoided? Let us compare the situation in Kyiv and in Hong Kong. These cities have similar population, but Hong Kong has introduced the relevant measures on December 31, 2019, as soon as alarming information from Wuhan appeared on a social media platform. The Hong Kong authorities immediately closed "transport ties with Wuhan" and increased "vigilance and temperature screenings at every border checkpoint, including the city's international airport and high-speed railway station in West Kowloon". The hospitals were told "to report further cases of "pneumonia of unknown origin" [91]. As a result, the accumulated number of cases in Hong Kong was 1045 on May 9, 2020. For example, the number of infected people in Kyiv was 1.7 times higher.

Quarantine has been announced all over Ukraine on March 12, 2020. It happened 9 days after the first COVID-19 case was confirmed in Chernivtsi region and 7 days before the first case in Kyiv. On May 11, the corresponding numbers of cases were 2396 and 1930, respectively. Taking into account that the population of Kyiv exceeds the population of Chernivtsi region more than three times, it can be concluded that the timely introduction of quarantine can effectively slow down spread of the infection.

6.13 The World

In this section, we consider the first wave of the COVID-19 pandemic and present the results of two SIR simulations for global dynamics.

SIR simulations for the COVID-19 epidemic dynamics in the world were performed with the use of V_j and t_j data sets presented in Tables 6.3 and 6.22 and two different time periods T_c taken for calculations. The results are shown in Table 6.25 (time moments correspond to Table 6.22) and Figs. 6.6, 6.7 and 6.29; some of them were published in [75, 77, 89]. The values of N, v, α, t_1^*, V_∞, t_{final} and other characteristics of the pandemic are very different for predictions 1 and 2. For example, the average time of spreading infection $1/\rho$ is 12 and 3.6 days for the first and second prediction, respectively. These figures are comparable with ones obtained for Ukraine and Moldova, but much higher than in the EU countries (0.77–1.47), USA (1.45), and the UK (3.03).

The basic reproduction number R_0 can be calculated with the use of formula (4.33) and the estimation of the doubling time (τ_d = 3.65 days) obtained in Chap. 2 with the use of the assumption that the number of cases increased exponentially. The exponential growth is characteristic only of the initial periods of an epidemic, so we can take the value of $\tau = 1/\rho = 12.035$ days, which corresponds to the prediction 1 (see Table 6.25) obtained using a data set for an earlier period. Then, Eq. (4.33) yields the value $R_0 = 9.83$ which looks too large.

The basic reproduction numbers of the COVID-19 pandemic were estimated in [12–14, 22, 30]. According to WHO calculations for China [12], values R_0 vary from 2 to 2.5. During the initial period of an epidemic outbreak, many infected persons are undetected. This fact diminishes the accuracy of R_0 estimations. For example, paper [30] reported the $R_0 = 7.97$ for Indonesia and two different values 1.54 and 6.22 for Singapore. If we use the estimation $\tau = 1/\rho = 3.636$ days

Table 6.25 First wave of the COVID-19 epidemic in the world

Number of prediction/optimal values of parameters, characteristics	Prediction 1 $n = 21$ Period taken for calculations T_c: April 9–29, 2020	Prediction 2 $n = 14$ Period taken for calculations T_c: April 29–May 12, 2020
N	6,637,317.12	23,600,000
v	3,537,150.79869746	19,925,870.4384
α	2.34901375364e-08	1.380196708625e-08
t_1^*	−111.265486657790	−170.837434239689
ρ	0.0830881587484244	0.275016207955738
$1/\rho$	12.0354093178044	3.63614932891861
r	0.999861835424859	0.999848053500794
F, Eq. (2.9)	68,744.3303788117	39,478.5832347979
$F/F_C(1, n-2)$	4522.65331439551	2122.50447498914
S_∞, Eq. (4.9)	1,595,881	16,654,425
V_∞, Eq. (4.16)	5,041,436	6,945,575
t_{final}, Eq. (4.17)	393	415.5
Final day of the first wave	February 16, 2021	March 11, 2021
V_f [86]	>32,730,945	>32,730,945
ε, Eq. (4.20)	>84.6%	>78.8%

Optimal values of parameters and other SIR model characteristics for different predictions

calculated in the second prediction (see Table 6.25), the corresponding global figure is $R_0 = 1.99$.

Unfortunately, we could correctly apply formula (4.20) only in February–March 2021 (after moments of time corresponding to t_{final}), but as early as September 26, 2020, the number of registered cases in the world 32,730,945 (see, e.g., [86]) already significantly exceeded theoretical estimates V_∞. In Table 6.25, we have used the inequality $V_f > 32{,}730{,}945$ to calculate the lower limit of possible values of relative error. It can be seen that the accuracy of SIR simulations of the global first wave (ε will exceed 79–85%) is the lowest one (in contrast to the situation for the EU countries, and even to USA, the UK, Ukraine, and the Republic of Moldova). So, the second and subsequent global waves of the COVID-19 pandemic began before the end of the first wave. We will discuss this in detail in the following chapters.

The SIR curves corresponding to prediction 1 are shown in Figs. 6.6 and 6.7. The epidemic dynamics in the world (brown lines and markers) was compared with the situation in USA (blue), Italy (green), Spain (yellow), the UK (red), and South Korea (magenta). The markers represent V_j values taken for calculations ("circles"), comparisons ("triangles") and verifications of calculations ("stars"). The number of laboratory confirmed cases in Wuhan, China, is shown by brown "squares." These values were calculated in [69] with the use of information from [6]. Solid brown line and brown "triangles" in Fig. 6.6 illustrate a big difference between number of calculated and actual cases which can be explained by the fact that many infected people do not have symptoms and there were and still are problems with their detection. In particular, the hidden periods (time between the estimated epidemic beginning t_1^* and the first confirmed case) can be rather long.

The value t_1^* estimated in prediction 1 (the beginning of the epidemic in late September 2019) does not look very reliable, since the registered number of global cases V_j was higher than the theoretical estimation $V(t)$ in February 2020 (compare brown "triangles" and brown solid line in Fig. 6.6). The above examples of SIR simulations in different countries show the opposite picture. Therefore, the estimate of the value of t_1^* obtained in prediction 2 is more credible, because all values V_j are below the corresponding theoretical curve $V(t)$ (see Fig. 6.29). The beginning of the pandemic should be sought in early August 2019, and in Ukraine it probably began in early 2020 (according to forecast 8, the results of which are also shown in Figs. 6.29 and 6.30).

Dashed magenta curve in Fig. 6.29 illustrates that more than 2000 persons could spread the infection on December 31, 2019—the day when China notified WHO about the situation in Wuhan. On January 23, 2020, this city was locked down, but the number of infected persons could be estimated by 6000 with hundreds of cases in USA, Italy, and Germany (see Fig. 6.6; the officially confirmed number of cases was 830 in mainland China, 2 in the Republic of Korea, and 1 in USA on this day). The Military World Games is likely to have played a fatal role in the spread of COVID-19 infection around the world. These competitions were held in Wuhan from October 18 to November 3, 2019, with the participation of 9300 athletes from

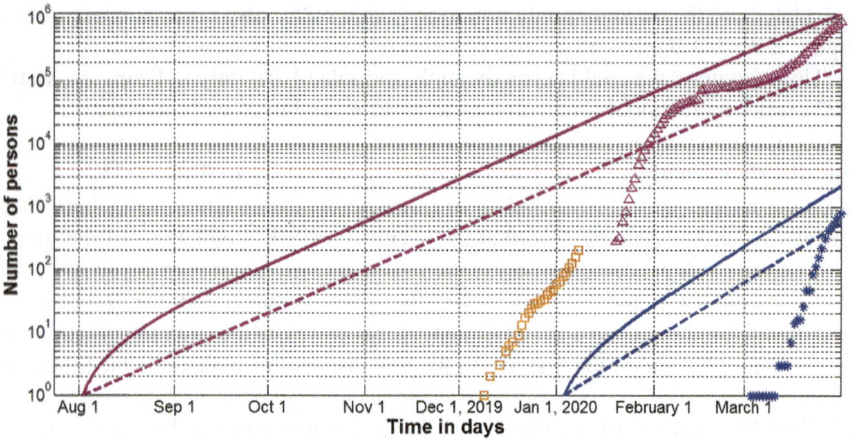

Fig. 6.29 World (prediction 2, magenta) and Ukraine (prediction 8, blue): SIR curves (lines) and accumulated number of cases (markers) versus time before March 1, 2020. Numbers of infected I (dashed) and the numbers of victims $V = I + R$ (solid); "triangles" correspond to the cases during initial stage of the epidemic; "stars"—the accumulated number of cases in Ukraine. The numbers of laboratory confirmed cases in Wuhan, China are shown by brown "squares"

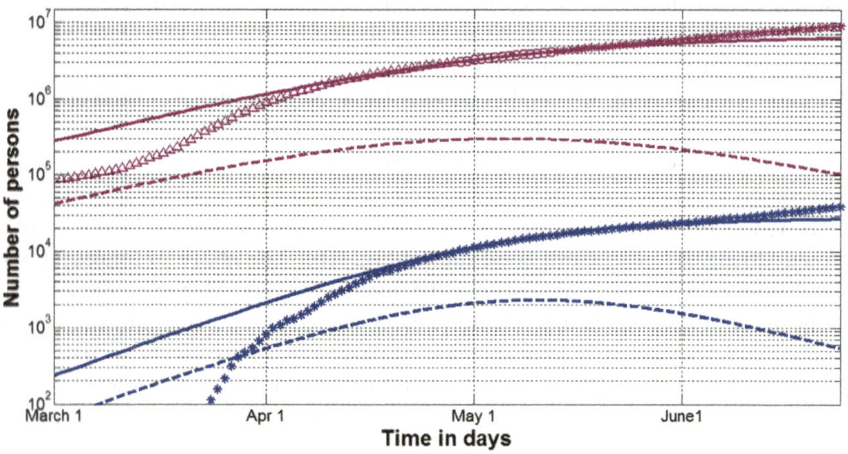

Fig. 6.30 World (prediction 2, magenta) and Ukraine (prediction 8, blue): SIR curves (lines) and accumulated number of cases (markers) versus time after March 1, 2020. Numbers of infected I (dashed) and the numbers of victims $V = I + R$ (solid); "circles" show the cases taken for calculations; "triangles" correspond to the cases during initial stage of the pandemic; "stars"—the data points used for a verification of the predictions

more than 140 countries. There are some reports that many participants became ill with COVID-19 symptoms and passed the infection onto their families [92].

Figure 6.30 shows that after June 1, 2020, the registered numbers of cases were much higher than the theoretical predictions both in Ukraine and in the world (compare "stars" and solid lines). This indicates the beginning of the second wave of the pandemic. In the following chapters, we will offer methods for detecting and modeling this and subsequent waves and give examples of calculations.

Chapter 7
Comparison of the First Waves of the COVID-19 Pandemic in Different Countries and Regions

In this chapter, we will compare and discuss the characteristics of the first waves of the COVID-19 pandemic in mainland China, USA, Germany, the UK, the Republic of Korea, Italy, Austria, Spain, France, the Republic of Moldova, Ukraine, the city of Kyiv and in the world, calculated in the previous chapter. Some results were published in [75, 77, 89]. The pandemic dynamics in different WHO regions will be discussed.

Some results of SIR simulations performed in previous chapter are summarized in Table 7.1. We have used the predictions corresponding to the smallest value of the relative error ε (see Eq. (4.20)) or the latest period T_c (the corresponding the data sets V_j and t_j were taken for calculations). In this chapter, we will use the characteristics introduced in previous Chapters. In particular, V_f is the real accumulated number of cases registered on the moment of time corresponding to the final day of the first wave t_{final} (Eq. (4.17)); the duration of the epidemics in days $T_d = t_{\text{final}} - t_1^*$ (Eq. 4.24); the estimation of the average time of spreading the infection $\tau = 1/\rho$ (in days); saturation level (or final size of the first wave) V_∞ (Eq. (4.16). For comparisons, we will use the accumulated number of COVID-19 cases V_A registered on September 26, 2020, [86]. In Table 7.1, the volume of population N_{pop} and the values of parameter N are shown in millions; values V_∞, V_f and V_A in thousands; α in $(\text{day})^{-1}$.

Table 7.1 and Fig. 7.1 demonstrate the absolute values of parameters taken from Chap. 6 and Internet (for the volumes of population N_{pop} in different countries and regions). In Fig. 7.1, the dependences α ("stars") and V_∞ ("circles") versus values of N_{pop} are shown (data points for China and South Korea are shown by "triangles"). It is seen that final size of the first wave V_∞ increases with increasing N_{pop} and the parameter α decreases. In the logarithmic scale, both dependences are close to linear (especially if we remove data points for China and South Korea).

We can try to find the best fitting lines for dependences

$$\alpha = \sigma N_{\text{pop}}^{\chi} \tag{7.1}$$

I. Nesteruk, *COVID-19 Pandemic Dynamics*,
https://doi.org/10.1007/978-981-33-6416-5_7

Table 7.1 Epidemic characteristics for different countries, regions and predictions calculated in Chap. 6. Absolute values

Country or region, number of prediction	N_{pop}	N, Eq. (4.5)	α (day)$^{-1}$ Eq. (4.1)	τ (days) Eq. (4.4)	T_d (days) Eq. (4.24)	V_∞, Eq. (4.16)	V_f	V_A
World, 2	7,794	23.6	1.38e-08	3.64	586.3	6,945.6	>3,2730.1	32,730.1
China, 2	1,441	0.14	4.28e-06	2.55	151.2	81.3	84.5	90.966
South Korea, 2	51.27	0.0137	1.92e-05	7.57	183.3	10.842	12.8	23.611
Ukraine, 8	43.72	0.059	4.62e-06	4.99	281.4	28.327	265.4	198.634
USA	331.0	7.082	1.09e-07	1.45	299	1,481.2	6,789.8	6,960.2
Germany, 2	83.78	0.882	9.37e-07	1.35	239	176.95	212.02	284.14
UK	67.89	0.480	9.14e-07	3.03	227	214.9	322.56	429.281
Italy, 4	60.46	1.13	6.76e-07	1.47	266	225.7	255.5	308.104
Spain, 2	46.75	0.76	1.07e-06	1.46	179	221.0	252.1	716.481
France, 2	65.28	0.69	1.97e-06	0.82	148	129.8	154.6	503.658
Austria, 6	9.01	0.0752	1.92e-05	0.77	102.3	15.108	16.179	42.940
Moldova, 2	4.03	0.00608	7.64e-05	3.23	110	3.541	12.254	50.534
Kyiv, 2	2.99	0.00303	1.19e-4	4.45	122	1.947	5.347	21.815

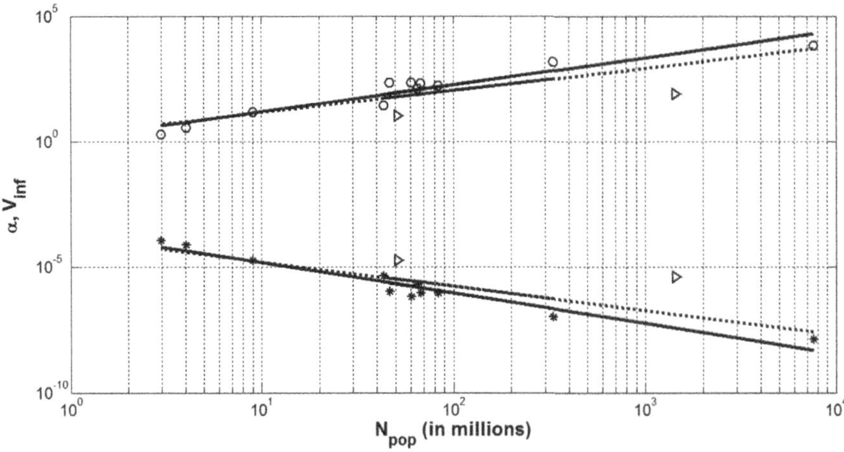

Fig. 7.1 Values of parameters α (in $(\mathrm{day})^{-1}$) and V_∞ (in thousands) versus volume of population. The data points from Table 7.1 for different countries and regions are shown by "stars" and "circles" for α and V_∞, respectively: "Triangles" represent data for mainland China and the Republic of Korea. The best fitting lines are dashed for complete data sets and solid for the data sets without China and Korea

$$V_\infty = \delta N_{\mathrm{pop}}^{\eta} \tag{7.2}$$

which correspond to the straight lines in logarithmic scale. For this purpose, we can use the method presented in Chap. 2 and Eq. (2.3), if we substitute y by $\ln \alpha$ and $\ln V_\infty$ for (7.1) and (7.2), respectively. Accordingly, the variable t has to be substituted by $\ln N_{\mathrm{pop}}$; the parameters γ have to be substituted by χ and η for (7.1) and (7.2), respectively; the parameter β—by $\ln \sigma$ and $\ln \delta$ for (7.1) and (7.2), respectively. Then we can calculate the values y_j and t_j with the use of data set presented in Table 7.1 and the linear regression (e.g., formulas (2.6)–(2.9)). The results of calculations are shown in Table 7.2 for different amount of elements in the data sets (n = 13 for complete data set and n = 11 for the data set without figures for China and Korea). Corresponding straight lines are shown in Fig. 7.1 (solid for n = 11 and dashed for n = 13)

Table 7.2 shows that both complete (n = 13) and incomplete (n = 11) data sets from Table 7.1 support the hypotheses about functional connections (7.1) and (7.2), because $|r| \approx 1$ and $F/F_C > 1$. In particular, relative value of the parameter α can be approximated by the following relationships:

$$\bar{\alpha} \equiv \alpha N_{\mathrm{pop}} = 0.0001466 N_{\mathrm{pop}}^{0.0375} \tag{7.3}$$

Table 7.2 Optimal values of parameters in Eqs. (7.1) and (7.2), correlation coefficients (2.7) and the results of Fisher test application

No.	No. of Eq.	n	Correlation coefficient r, Eq. (2.8)	Optimal values of parameters χ or η in (7.1) or (7.2)	Optimal values of parameters α or δ in (7.1) or (7.2)	Experimental value of the Fisher function F, Eq. (2.9), $m = 2$	Critical value of Fisher Function $F_c(1, n-2)$ for the confidence level 0.001, [34]	F/F_c
1.	(7.2)	13	0.8368	0.8900	1.8546	25.6897	20.3	1.27
2.	(7.1)	13	−0.8240	−0.9625	0.0001466	23.2684	20.3	1.15
3.	(7.2)	11	0.9532	1.0778	1.3101	89.5132	22.9	3.91
4.	(7.1)	11	−0.9692	−1.2049	0.0002401	139.1859	22.9	6.08

or

$$\bar{\alpha} \equiv \alpha N_{\text{pop}} = 0.0002401 N_{\text{pop}}^{-0.2049} \tag{7.4}$$

The relative final size of the first epidemic wave V_∞ can be approximated by following equations:

$$\bar{V}_\infty \equiv \frac{V_\infty}{N_{\text{pop}}} = 0.0018546 N_{\text{pop}}^{-0.11} \tag{7.5}$$

$$\bar{V}_\infty \equiv \frac{V_\infty}{N_{\text{pop}}} = 0.0013101 N_{\text{pop}}^{0.0778} \tag{7.6}$$

for complete ($n = 13$) and incomplete ($n = 11$) data sets, respectively.

The proposed relative values are very weakly dependent on the size of the population (see (7.3)–(7.6)), so they can be used in the system of differential Eqs. (4.1)–(4.3), which takes the following form:

$$\frac{d\bar{S}}{dt} = -\bar{\alpha}\bar{S}\bar{I} \tag{7.7}$$

$$\frac{d\bar{I}}{dt} = \bar{\alpha}\bar{S}\bar{I} - \rho\bar{I} \tag{7.8}$$

$$\frac{d\bar{R}}{dt} = \rho\bar{I} \tag{7.9}$$

where

$$\bar{I} \equiv \frac{I}{N_{\text{pop}}},\ \bar{S} \equiv \frac{S}{N_{\text{pop}}},\ \bar{R} \equiv \frac{R}{N_{\text{pop}}} \tag{7.10}$$

System of differential Eqs. (7.7)–(7.9) contains two unknown parameters $\bar{\alpha}$ and ρ, but it would be wrong to assume that its solution in relative quantities depends only on these two parameters, since initial conditions (4.6) contain an unknown ratio $\bar{N} = N/N_{\text{pop}}$. Really, Eq. (4.6) can be rewritten as follows:

$$\bar{I}(t_1^*) = 1/N_{\text{pop}},\ \bar{R}(t_1^*) = 0,\ \bar{S}(t_1^*) = \bar{N} - 1/N_{\text{pop}} \tag{7.11}$$

In particular, this ratio influences the value $\bar{S}_\infty \equiv S_\infty/N_{\text{pop}}$, which can be defined form the nonlinear equation

$$\bar{S}_\infty = \left(\bar{N} - \frac{1}{N_{\text{pop}}}\right)e^{\frac{\bar{S}_\infty - \bar{N}}{\bar{v}}},\quad \bar{v} = \frac{\rho}{\bar{\alpha}} \tag{7.12}$$

following from (4.9). After solving Eq. (7.12), the final size of the epidemic $\bar{V}_\infty$ can be calculated from

$$\bar{V}_\infty = \bar{N} - \bar{S}_\infty \tag{7.13}$$

Equation (7.13) follows from (4.5). Nonlinearity of Eq. (7.12) leads to complicated nonlinear approximations (7.5) or (7.6).

Another important epidemic characteristic—its duration must be calculated from nonlinear relationship:

$$T_d = -\frac{2}{\bar{\alpha}} \int\limits_{\bar{N}-1/N_{\text{pop}}}^{\bar{v}} \frac{d\bar{U}}{\bar{U}\left[\bar{v}\ln\bar{U} - \bar{U} + \bar{N} - \bar{v}\ln\left(\bar{N} - 1/N_{\text{pop}}\right)\right]} \tag{7.14}$$

which follows from (4.25). Equations (7.12)–(7.14) demonstrate that correct estimations of the epidemic parameters can be done only with the use of four unknown independent parameters N, v, α, t_1^* (for system (4.1)–(4.3)) or $\bar{N}$, $\bar{v}$, $\bar{\alpha}$, t_1^* (for system (7.7)–(7.9)). Dependences (7.5) and (7.6) can be treated as some approximations which allow us to estimate the final sizes of the first epidemic waves without identification of the optimal values of these parameters as was done in Chaps. 5 and 6. The accuracy of these approximations will be checked later in this chapter.

Figure 7.2 demonstrates the absolute values of parameters taken from Table 7.1 versus values of N. The values α are shown by "stars" and V_∞—by "circles"; data points for China and South Korea are represented by "triangles". It is seen that with increasing N, final size of the first wave V_∞ increases and the parameter α decreases. In the logarithmic scale, both dependences are very close to linear (including the data points for China and South Korea).

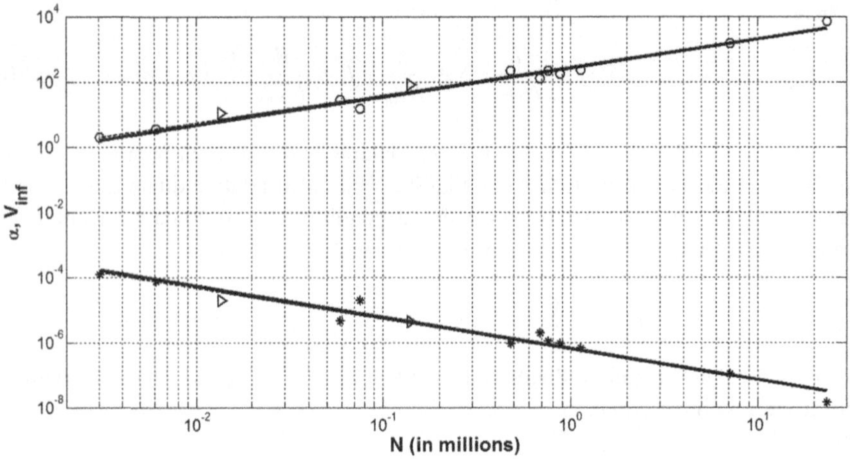

Fig. 7.2 Values of parameters α (in (day)$^{-1}$) and V_∞ (in thousands) versus values of SIR parameter N. The data points from Table 7.1 for different countries and regions are shown by "stars" and "circles" for α and V_∞, respectively: "Triangles" represent data for mainland China and the Republic of Korea. The best fitting lines are dashed for complete data sets and solid for the data sets without China and Korea.

We can try to find the best fitting lines for dependences

$$\alpha = \sigma N^{\chi} \tag{7.15}$$

$$V_\infty = \delta N^{\eta} \tag{7.16}$$

which correspond to the straight lines in the logarithmic scale. For this purpose, we can use the method presented in Chap. 2, if we substitute in (2.3) y by $\ln\alpha$ and $\ln V_\infty$ for (7.15) and (7.16), respectively. Accordingly, the variable t has to be substituted by $\ln N$; parameter γ—by χ and η for (7.15) and (7.16), respectively; parameter β—by $\ln\sigma$ and $\ln\delta$ for (7.15) and (7.16), respectively. Then we can calculate the values y_j and t_j with the use of data set presented in Table 7.1 and the linear regression (formulas (2.6)–(2.9)). The results of calculations are shown in Table 7.3 for different amount of elements in the data sets (n = 13 for complete data set and n = 11 for the data set without figures for China and Korea). Corresponding straight lines are shown in Fig. 7.2 (solid for n = 11 and dashed for n = 13).

Table 7.3 and Fig. 7.2 demonstrate that both complete (n = 13) and incomplete (n = 11) data sets from Table 7.1 support the hypotheses about functional connections (7.15) and (7.16). In comparison with Table 7.2, the values of parameters r, F, χ, η, σ and δ are very close and figures for $|r|$ and F/F_C are higher The difference in best fitting lines for complete (n = 13, dashed) and incomplete (n = 11, solid) is almost invisible in Fig. 7.2. The best approximations (corresponding to the highest values of F/F_C ratio) for the parameter α and V_∞(in thousands) can be approximated by the following relationships:

Table 7.3 Optimal values of parameters in Eqs. (7.15) and (7.16), correlation coefficients (2.7) and the results of Fisher test application

No.	No. of Eq.	n	Correlation coefficient r, Eq. (2.8)	Optimal values of parameters χ or η in (7.15) or (7.16)	Optimal values of parameters α or δ in (7.15) or (7.16)	Experimental value of the Fisher function F, Eq. (2.9), $m = 2$	Critical value of Fisher Function $F_c(1, n-2)$ for the confidence level 0.001, [34]	F/F_c
1.	(7.16)	13	0.9865	0.8598	281.862	400.119	20.3	19.7
2.	(7.15)	13	−0.9771	−0.9352	6.35645e-07	231.805	20.3	11.3
3.	(7.16)	11	0.9897	0.8829	267.228	431.728	22.9	18.9
4.	(7.15)	11	−0.9786	−0.9598	6.48634e-07	203.111	22.9	8.9

$$\alpha = 6.35645\text{e - }07 N^{-0.9352} \tag{7.17}$$

$$V_{\infty} = 400.119 N^{0.8598} \tag{7.18}$$

In relationships (7.17) and (7.18), the values of SIR parameter N must be taken in millions. These approximate formulas can be useful to control new SIR simulations due to the high levels of F/F_C ratio. But they contain a priori unknown parameter N (its values can be estimated with the use of the procedure proposed in Chap. 5). In comparison, formulas (7.3)–(7.6) allow estimating the values of epidemic parameters without any new SIR simulations. The relative characteristics based on known values N_{pop} and calculated with the use of data from Table 7.1 are presented in Table 7.4. Later we will show the corresponding characteristics for many other countries and compare with the estimations (7.5) and (7.6).

Table 7.4 demonstrates that the lowest values of V_{∞}/N_{pop} were predicted for China and South Korea. The first waves of epidemic in these countries have finished with figures of V_f/N_{pop} which are slightly higher than predicted (the relative error ε is low). As of September 26, 2020, the values V_A/N_{pop} did not sufficiently exceed the V_f/N_{pop} ones. It means that next waves of the pandemic in these two countries were very moderate. In Ukraine, Moldova, Kyiv and in the world, predicted values of V_{∞}/N_{pop} were much higher. The first waves of epidemic in these countries have finished (or even continue) with figures of V_f/N_{pop} which are much higher than predicted ones (the relative error ε is very high). It means that these countries probably had problems with detection and isolation of infected persons. As of September 26, 2020, the values V_A/N_{pop} dramatically exceeded the V_f/N_{pop} ones (prediction 8 for

Table 7.4 Relative epidemic characteristics for different countries, regions and predictions calculated in with the use of Table 7.1

Country or region, number of prediction	N/N_{pop}	$\alpha \cdot N_{pop}$ (day)$^{-1}$	V_{∞}/N_{pop}	V_f/N_{pop}	ε (%)	V_A/N_{pop}
World, 2	0.0031	104.8	0.000915	>0.00431	>78.8	0.00431
China, 2	9.715e-05	6.167	5.64e-05	5.86e-05	3.8	6.31e-05
South Korea, 2	0.000267	984.4	0.000211	0.000250	15.3	0.000461
Ukraine, 8	0.00135	202.0	0.000648	0.00595	89.3	0.00454
USA	0.02140	36.1	0.00448	0.0205	78.2	0.0210
Germany, 2	0.01053	78.5	0.00211	0.00253	16.5	0.00340
UK	0.00707	62.1	0.00317	0.00475	33.4	0.00632
Italy, 4	0.01869	40.9	0.00373	0.00423	11.7	0.00510
Spain, 2	0.01626	50.0	0.00473	0.00539	12.3	0.0153
France, 2	0.01057	128.6	0.00199	0.00237	16.0	0.00772
Austria, 6	0.00835	173.0	0.00168	0.00180	6.6	0.00477
Moldova, 2	0.00151	307.9	0.000879	0.00304	71.1	0.0125
Kyiv, 2	0.00101	355.8	0.000651	0.00179	63.6	0.00730

Ukraine looks like an exception, since its final day is October 11, 2020). In addition to problems with identifying patients, this may also be the result of premature easing of quarantine and mass non-compliance with social distance.

The highest values of V_∞/N_{pop} were predicted for the USA and Spain, but the first wave of epidemic in Spain has finished with small enough value of the relative error ε. It means that quarantine and social behavior in this country were effective enough to stabilize the first wave. We saw a different picture in the USA, which led to the beginning of the next waves before the end of the first one. This is evidenced by the high error ε and the largest V_A/N_{pop} value (among presented in Table 7.4).

Figures 7.3, 7.4 and 7.5 illustrate the information presented in Tables 7.1 and 7.4. It can be seen values of the relative error ε ("crosses x") do not depend on the volume of population, parameters $\alpha \cdot N_{\text{pop}}$ or τ. The highest values of error correspond to Ukraine, Moldova, Kyiv, USA, and the world yielding the large difference in values of V_∞/N_{pop} ("circles"), V_f/N_{pop} ("triangles") and V_A/N_{pop}("crosses +"). For low values of the relative error ε (mainland China, South Korea, Austria, Italy, Germany, France, and Spain), these differences are much smaller. The duration of the first wave varies from 102 to 586 days and is lower for countries and regions with smaller population. The estimation of the average time of infection spreading τ varies in the range 0.8–7.6 days and shows no signs of dependence on population size or parameter $\alpha \cdot N_{\text{pop}}$ (see "dots" in Figs. 7.3 and 7.4).

Figure 7.4 illustrates the values presented in Tables 7.1 and 7.4 versus relative value of the parameter α (i.e., the product $\alpha \cdot N_{\text{pop}}$; the points corresponding China and Korea are not shown). We cannot see any visible dependences of SIR parameters and other characteristics of the first COVID-19 pandemic wave on the

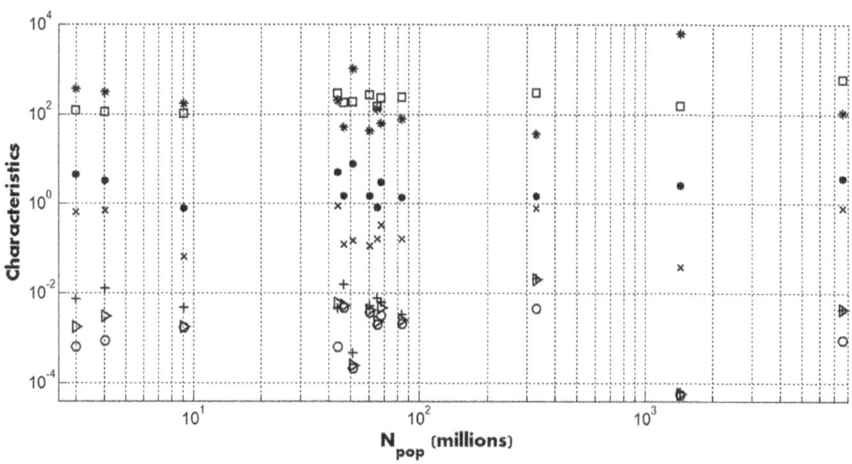

Fig. 7.3 Epidemic characteristics from Tables 7.1 and 7.4 versus volume of population. "Circles" represent V_∞/N_{pop}; "triangles"—V_f/N_{pop}; "crosses +"— V_A/N_{pop}; "crosses x"—the relative error ε; "dots"—the estimation of the average time of infection spreading τ (in days); "squares"—the first epidemic wave duration T_d (in days); "stars"—$\alpha \cdot N_{\text{pop}} \cdot 10^6$(the volume of population N_{pop} is taken in millions)

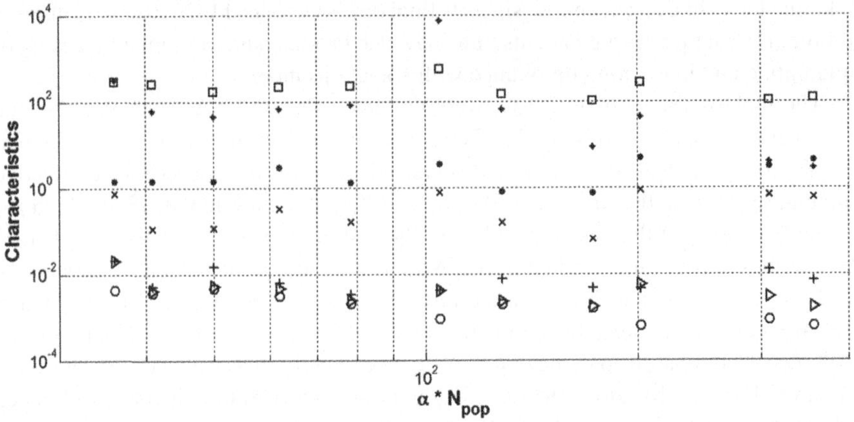

Fig. 7.4 Epidemic characteristics from Tables 7.1 and 7.4 versus relative value of the parameter α. "Circles" represent $V_{\infty}/N_{\text{pop}}$; "triangles"—$V_f/N_{\text{pop}}$; "crosses +"— V_A/N_{pop}; "crosses x"—the relative error ε; "dots"—the estimation of the average time of infection spreading τ (in days); "squares"—the first epidemic wave duration T_{d} (in days); "stars"—the volume of population N_{pop}(here in millions). Data for China and South Korea are not shown

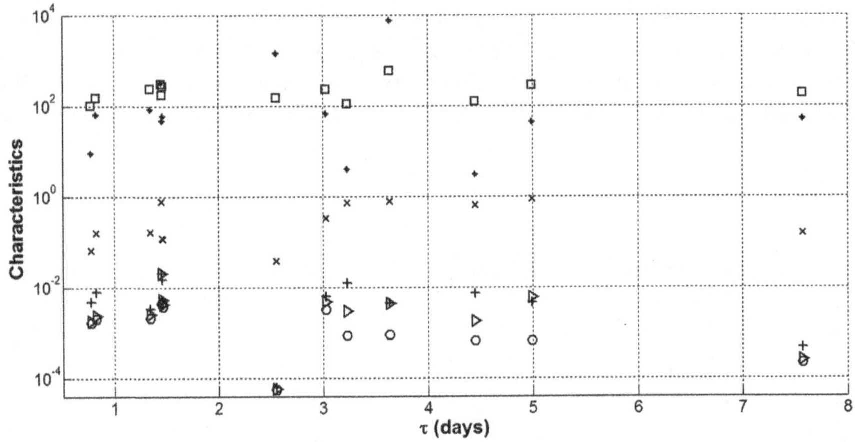

Fig. 7.5 Epidemic characteristics from Tables 7.1 and 7.4 versus the estimation of the spreading time. "Circles" represent $V_{\infty}/N_{\text{pop}}$; "triangles"—$V_f/N_{\text{pop}}$; "crosses +"—$V_A/N_{\text{pop}}$;"crosses x"—the relative error ε; "squares"—the first epidemic wave duration T_d (in days); "stars"—the volume of population N_{pop}(here in millions)

values of $\alpha \cdot N_{\text{pop}}$. The relative value of the parameter α calculated for mainland Chine and the Republic of Korea is much higher in comparison with other countries and regions (see Table 7.4). This fact needs a special discussion and investigation. We will consider the peculiarities of the COVID-19 pandemic in these countries and corresponding WHO region later.

Figure 7.5 illustrates the values presented in Tables 7.1 and 7.4 versus the estimation of the spreading time $\tau = 1/\rho$. It can be seen that the smallest values of V_∞/N_{pop} ("circles"), V_f/N_{pop} ("triangles") and V_A/N_{pop} ("crosses +") correspond to the cases of China and Korea. This fact again draws special attention to the dynamics of the pandemic in these countries.

Formulas (7.5) and (7.6) can be used to estimate the final sizes of the first pandemic waves in different WHO regions. But first, we need to analyze how the pandemic waves look. We can use the information from the WHO Weekly Epidemiology Update [93] where the number of new COVID-19 cases confirmed every week are shown. For example, the global situation is shown in Fig. 7.6. Figures 7.7, 7.8, 7.9, 7.10, 7.11 and 7.12 illustrate the number of new cases reported every week for six different WHO regions (Africa, Americas, Eastern Mediterranean, Europe, South-East Asia, and Western Pacific).

If we denote the accumulated number of cases confirmed at the end of i-th week as w_i (do not confuse with the daily accumulated number of cases V_j), the diagrams presented in Figs. 7.6, 7.7, 7.8, 7.9, 7.10, 7.11 and 7.12 show the values of $w_i - w_{i-1}$. In terms of SIR model, these values can be treated as estimations of dV/dt:

$$w_i - w_{i-1} \approx 7\mathrm{d}V/\mathrm{d}t \tag{7.19}$$

Coefficient "7" appeared in formula (7.19) since we measure time in days for the theory presented in Chap. 4. Then Eq. (4.11) yields the following approximation:

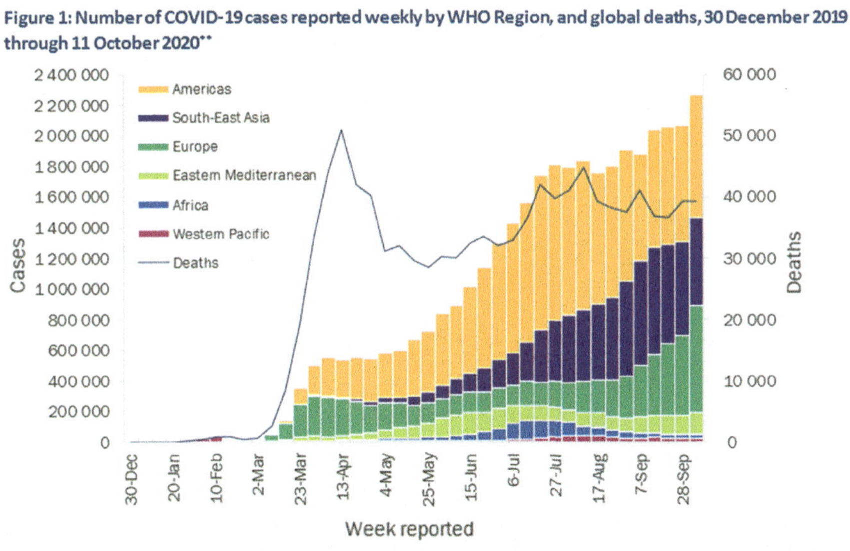

Fig. 7.6 Global number of new COVID-19 cases reported every week, [93]

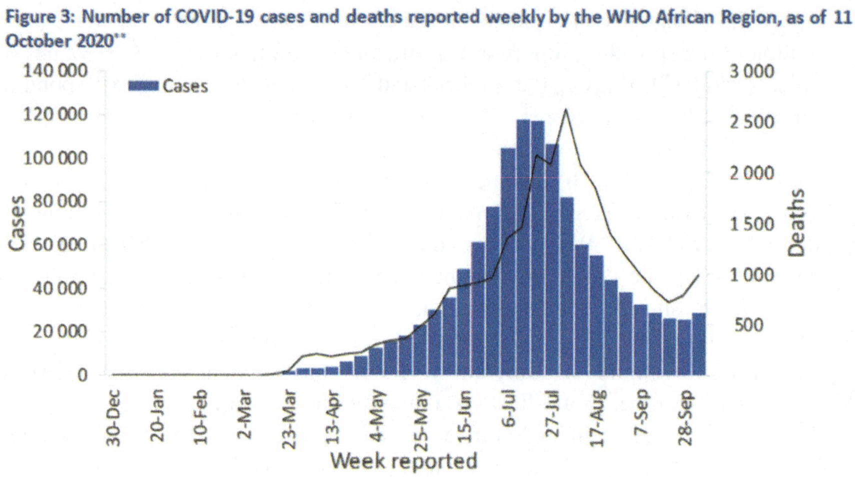

Fig. 7.7 Number of new COVID-19 cases reported every week by the WHO African region, [93]

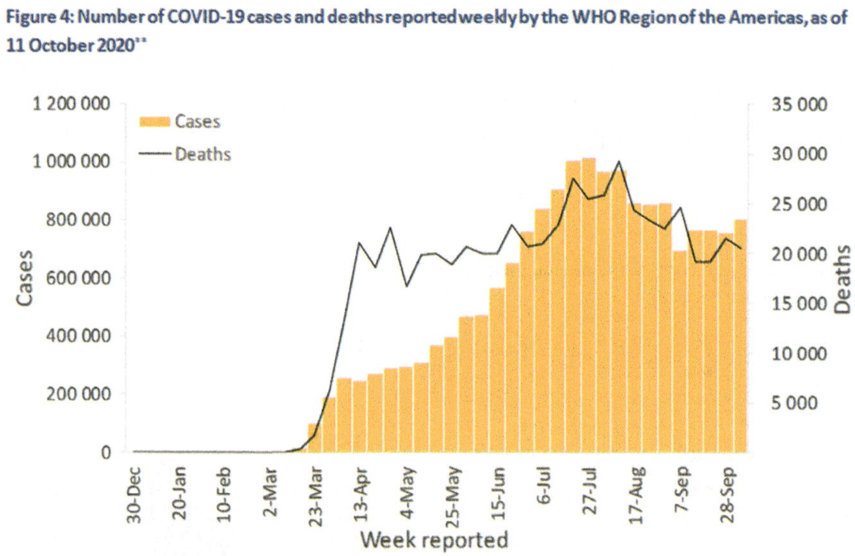

Fig. 7.8 Number of new COVID-19 cases reported every week by the WHO region of the Americas, [93]

$$w_i - w_{i-1} \approx 7\alpha SI \tag{7.20}$$

At the final moment of the first pandemic wave $t = t_{\text{final}}$, Eq. (7.20) yields very small values of $w_i - w_{i-1}$ since $S \rightarrow S_\infty$ and $I \rightarrow 0$ at infinity. Equations (4.9) and (4.17) allow us to estimate the corresponding $w_i - w_{i-1}$ value more precisely:

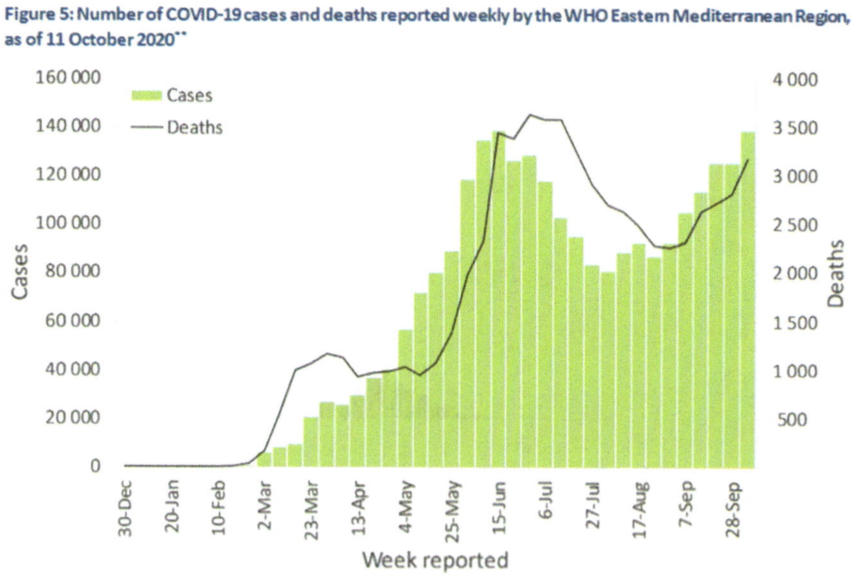

Fig. 7.9 Number of new COVID-19 cases reported every week by the WHO Eastern Mediterranean region, [93]

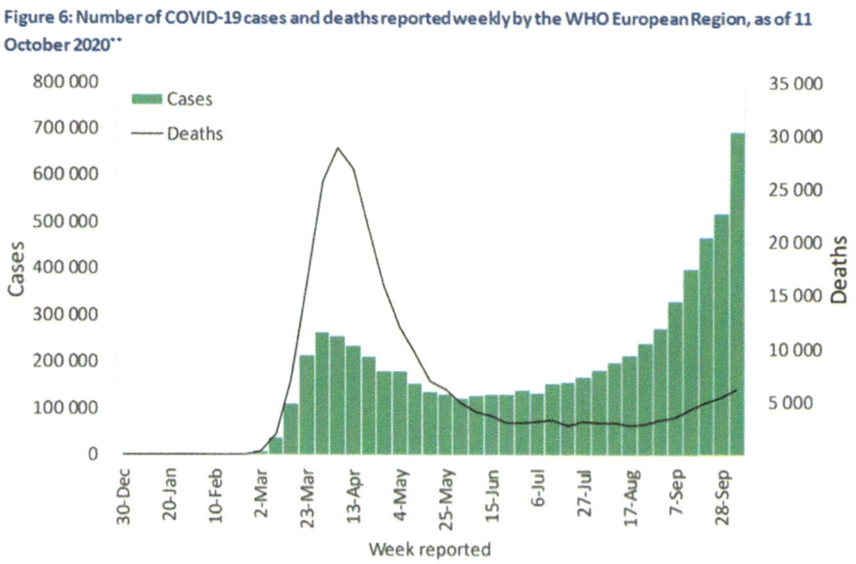

Fig. 7.10 Number of new COVID-19 cases reported every week by the WHO European region, [93]

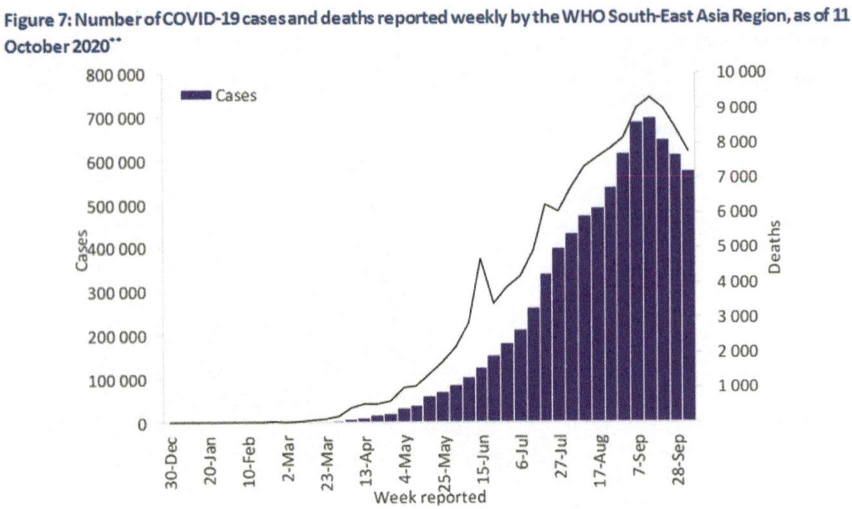

Fig. 7.11 Number of new COVID-19 cases reported every week by the WHO South-East Asia region, [93]

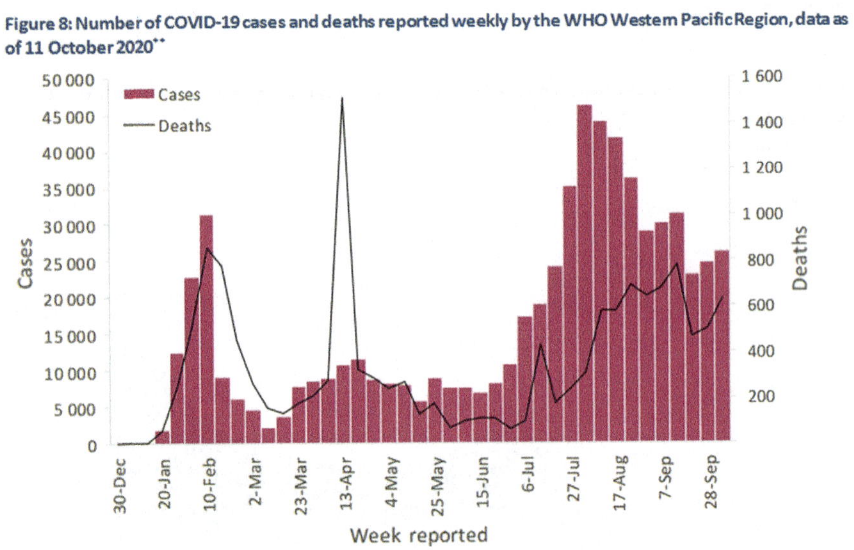

Fig. 7.12 Number of new COVID-19 cases reported every week by the WHO Western Pacific region, [93]

$$(w_i - w_{i-1})|_{t=t_{\text{final}}} \approx 7\alpha S_\infty \tag{7.21}$$

To calculate the right part of Eq. (7.21), we can use the values of SIR parameters presented in Table 6.25 (second prediction for the world). The result $(w_i - w_{i-1})|_{t=t_{\text{final}}} \approx 1.6$ shows that first wave of the pandemic must end with almost zero values of $w_i - w_{i-1}$.

Almost zero weekly (and daily) increase in the global number of COVID-19 cases was observed in late February 2020. It is difficult to see it in diagram 7.6, but clearly visible in Fig. 6.29. We cannot treat this period as the end of the first pandemic wave, since too many cases in many countries were not detected in this period (compare magenta solid line and "triangles" in Fig. 6.29). After March 2, 2020, the values $w_i - w_{i-1}$ increased almost monotonically. A slight minimum is visible only around August 17, 2020 (see Fig. 7.6). It means that next waves of pandemic started before the first waves ended. We will discuss this issue in details later.

Here we can try to use the time moment t_B corresponding the minimum of $w_i - w_{i-1}$ as an estimation of the last day of the global first wave and for the WHO regions as well. The values t_B corresponding to the accumulated number of cases V_B are listed in Table 7.5. It can be seen that global value of V_B exceeds the theoretical estimation V_∞ (see Table 7.1) almost three times. Let us analyze the situation in individual regions.

Figure 7.7 illustrates that the first wave of the epidemic probably ended in late September 2020. Compared with the global dynamics, the minimum of values $w_i - w_{i-1}$ is much lower than the maximum figures. The first wave of the pandemic in the WHO African Region is much more pronounced than in the world (compare Figs. 7.6 and 7.7), but the next (or subsequent) pandemic wave/waves in Africa also began before the end of the first one.

Figure 7.8 illustrates that the minimum of $w_i - w_{i-1}$ values corresponds to the beginning of September 2020 and is not much lower than the maximum figures corresponding to mid-July. The first wave of the pandemic in the WHO region of the Americas is much less pronounced than in Africa. Probably in the Americas, some subsequent pandemic waves began before the end of the first one. This is also evidenced by much higher values of V_B/N_{pop} compared to other regions and the world as a whole (see Table 7.5).

Figure 7.9 illustrates that the minimum of $w_i - w_{i-1}$ values corresponds to the beginning of August 2020 and is rather lower than the maximum figures corresponding to mid-June, but the first wave of the pandemic in the WHO Eastern Mediterranean region is less pronounced than in Africa. Probably in Mediterranean, some subsequent pandemic waves began before the end of the first one.

Figure 7.10 illustrates that the minimum of $w_i - w_{i-1}$ values corresponds to rather long period in June 2020. This fact is in good agreement with the calculations for the last day of the first wave performed in the previous chapter for many European countries and summarized in Table 7.6 (see also [75, 77]). This table shows not only predictions for the last day of the first epidemic wave t_{final}, but also

Table 7.5 COVID-19 pandemic characteristics and theoretical estimations for different WHO regions

WHO region	N_{pop} (millions), [93]	V_A	V_A/N_{pop}	Last day of the first wave in 2020 t_B	$V_B = V_j(t_B)$	V_B/N_{pop}	V_∞/N_{pop} Eq. (7.5)	V_∞/N_{pop} Eq. (7.6)
Africa	1,122.2	1,172,342	1,045	September 26	1,172,342	1,045	856	2,262
Americas	1,022.7	16,233,110	15,873	September 5	14,001,390	13,691	865	2,246
Eastern Mediterranean	730.8	2,340,215	3,202	August 2	1,564,836	2,141	897	2,188
Europe	933.4	5,662,875	6,067	May 31	2,142,547	2,295	873	2,230
South-East Asia	2,021.2	6,720,771	3,325	After October 3	7,911,036	3,914	802	2,368
Western Pacific	1,963.4	600,891	306	March 7	88,787	45	805	2,363
Grand total (the world)	7,794	32,730 945	4,199	August 15	21,294,845	2,732	692	2,630

its duration $t_{\text{final}} - t_1^*$ and the hidden period (the number of days between the estimated epidemic outbreak t_1^* and the day when the first COVID-19 case was confirmed). It can be seen that the longest epidemic durations were expected in the countries with the longest hidden periods (USA, Italy and Germany). After June 2020, values $w_i - w_{i-1}$ started to increase in the WHO European region. This growth became very rapid in September–October, indicating the intense subsequent waves of the pandemic in Europe. Next we will dwell in detail on the course of these waves in Ukraine and predict the final sizes of new epidemic waves in USA, Germany, the UK, the Republic of Korea, Italy, Austria, Spain, France, Sweden, the Republic of Moldova, Ukraine, the city of Kyiv, and the whole world.

Figure 7.11 illustrates that the minimum of $w_i - w_{i-1}$ values probably was not achieved in the beginning of October 2020. We may expect the first pandemic wave did not finish in the WHO South-East Asia region. For comparison with other regions, we take the accumulated number of cases V_B corresponding to October 3, 2020 (see Table 7.5). Unfortunately, the value $V_B/N_{\text{pop}} = 3914$ was already much higher than the global figure 2732 and the corresponding ratios in other regions (an exception is only the Americas, see Table 7.5).

Table 7.6 First waves of the COVID-19 pandemic. Epidemic characteristics for different countries and regions. Estimated days of epidemic outbreak, hidden periods, epidemic durations, the final days and number of cases

Country or region, number of prediction	Last day of data set used for calculations	Estimated day of epidemic outbreak, t_1^*	Hidden period, days	Epidemic duration, $t_{\text{final}} - t_1^*$, days	Last day of epidemic, t_{final}	Final number of cases, V_∞
World, 1	April 29	October 1	68	537	March 21, 2021	5,041,436
USA	April 29	November 28	55	300	September 23	1,481,153
Germany, 2	April 29	December 10	48	238	August 4	176,952
UK	April 29	January 6	25	227	August 20	214,899
South Korea, 2	April 29	December 28	23	183	June 29	10,842
Italy, 4	April 18	November 26	65	266	August 18	225,736
Spain, 2	April 18	January 9	22	179	July 6	221,011
France, 2	April 18	January 24	0	148	June 20	129,823
Austria, 6	April 10	February 6	19	103	May 19	15,108
Moldova, 2	April 18	February 26	10	110	June 15	3,541
Ukraine, 3	April 24	February 16	15	160	July 26	15,273
Kyiv, 2	April 24	March 2	17	122	July 2	1,947

Figure 7.12 shows two distinct minima of $w_i - w_{i-1}$ distribution, corresponding to the beginning of March and the middle of May 2020. There are also some local minima, indicating rather irregular statistics in the WHO Western Pacific region. Another characteristic feature of the pandemic dynamics in this region is the very low number of cases per capita. The value $V_B/N_{pop} = 45$ is much lower than the global figure 2732 and the corresponding ratios in other regions (see Table 7.5). To compare the number of cases per capita corresponding to the fixed day, the accumulated number V_A registered on September 26, 2020, [86] is also shown in Table 7.5. The value $V_A/N_{pop} = 306$ much lower than the global figure 4199 and the corresponding ratios in other regions (see Table 7.5 and compare magenta and black markers in Fig. 7.13).

Figure 7.13 shows that the statistical estimation (7.6) agrees quite well with the V_B/N_{pop} values ("triangles") for the world and many regions. But the real numbers of cases per capita were higher in the Americas (orange markers), lower in Africa (blue) and much lower in the WHO Western Pacific region (magenta). Estimation (7.5) agrees only with the data for the WHO African region. It is worth recalling that Eq. (7.6) corresponds to the best fitting line for SIR simulations data without taking into account China and Korea (n = 11), which are in the Western Pacific region, and Eq. (7.5) taking into account them (n = 13). The obtained results testify to the special nature of the pandemic dynamics in this region. Explanations of this fact are still waiting for its researchers, but let us express some hypotheses.

First of all, the pandemic probably started in this region in August 2019 (according to the estimation of t_1^* for the world, see Table 6.25, prediction 2). It means that first COVID-19 cases in China (probably in South Korea and other countries)

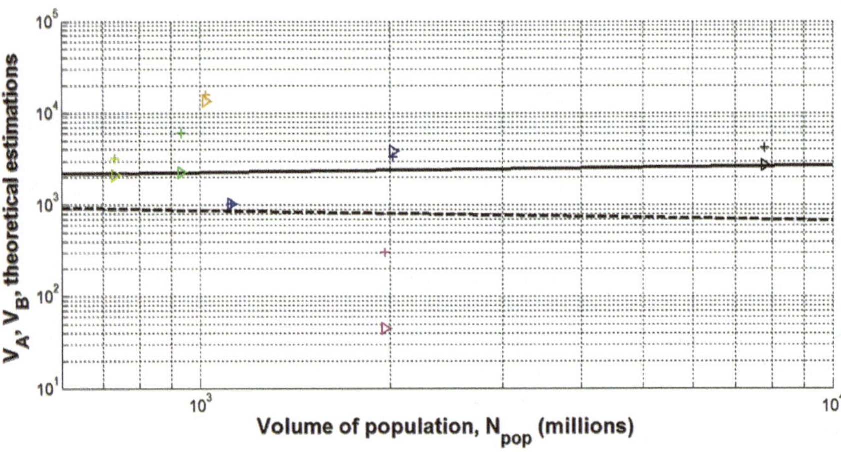

Fig. 7.13 Registered number of cases per capita in different WHO regions and theoretical estimations versus the volume of population. "Crosses" represent V_A/N_{pop} values: "triangles"—V_B/N_{pop}; colors of markers are similar to those used in Figs. 7.6, 7.7, 7.8, 7.9, 7.10, 7.11 and 7.12 for different WHO regions, global values are shown in black; the volume of population N_{pop} is measured in millions. Dashed line corresponds to estimation (7.5); solid one to (7.6)

were not identified and registered during at least 4 months. Probably, these first cases were not very severe, and the symptoms were not so pronounced. Presumably mutations of the coronavirus made it more pathogenic, and sick people became more noticeable in December 2019. But previous cases were not taken into account in the statistics. It should not be ruled out that the population of this region had a collective immunity to pathogens similar to COVID-19 before the pandemic.

We do not question the effectiveness of quarantine measures and the speed of isolation of patients and contacts in China, South Korea, Japan, Australia or New Zealand. But it is difficult to explain the low incidence per capita in the region only by this effectiveness. Probably, some kind of a collective immunity has the African population as well; since the number of cases per capita is significantly lower than in other regions (except Western Pacific). High values of V_A/N_{pop} and V_B/N_{pop} for the WHO region of Americas may indicate insufficient quarantine and its massive violation (compared to Europe and Mediterranean).

Chapter 8
Identification of the New Waves of the COVID-19 Pandemic

A simple method for the detection of epidemic waves will be proposed and applied for the COVID-19 pandemic dynamics in USA, Germany, the UK, the Republic of Korea, Italy, Austria, Spain, France, Sweden, the Republic of Moldova, Ukraine, the city of Kyiv, and in the world.

In previous chapter, we have already discussed the next waves of the pandemic which started before the first waves ended. Constant changes in the pandemic conditions (in particular, the peculiarities of quarantine and its violation, situations with testing and isolation of patients) cause the changes in the values of parameters of the mathematical models. In this chapter, we develop simple methods of detecting these changes, and in the next chapters we will try to improve the mathematical simulations taking into account changeable parameters. Some results have already been published in [78, 89, 94].

The SIR curves (e.g., calculated in Chap. 6 for the first wave of epidemics) reflect the situation inherent in the period of time used for calculations. Let us discuss first the changes in the values of SIR parameters which can occur due to the quarantine relaxations. First of all, more contacts lead to increase in the number of susceptible persons, i.e., new value S_N could be much higher than S during lockdown. SIR model is able to estimate corresponding consequences. Thus, the daily number of new cases can be estimated by the derivative dV/dt, which increases after releasing the quarantine according to Eq. (4.11), since $S_N > S$. For example, we can see this in Fig. 6.16. In Austria, small shops were opened after April 14, 2020. The confirmed accumulated number of cases (red "stars") deviates form the red solid SIR curve after the relevant incubation period.

If the increase in the value of the parameter S_N is compensated by a decrease in the number of non-isolated infected persons I, the right-hand side of Eq. (4.11) may be close to a constant value. This situation corresponds to an approximately linear increase in the number of cases. We observe such a situation in Austria in May–June 2020 (see red "stars" in Fig. 6.16). Unfortunately, the number of cases in

I. Nesteruk, *COVID-19 Pandemic Dynamics*,
https://doi.org/10.1007/978-981-33-6416-5_8

Ukraine increased much more intensively (see blue "stars" in Fig. 6.16). Probably, this indicates that the cancelation of the national lockdown was carried out too early. May 10, 2020, corresponded to the maximum of non-isolated infected persons in Ukraine (see the blue dotted line in Fig. 6.16). The violation of quarantine by Ukrainian authorities and population caused much more intensive increase in the number of cases than in Austria.

To control the changes of epidemic parameters, we can use daily or weekly numbers of new cases and their derivatives. Since these values are random, we need some smoothing (especially for daily amounts, which are also characterized by some weekly periodicity). For example, we can use the smoothed daily number of accumulated cases:

$$\bar{V}_i = \frac{1}{7}\sum_{j=i-3}^{j=i+3} V_j, \tag{8.1}$$

or weekly number of cases w_i (WHO reported only these values after August 15). The first and second derivatives can be estimated with the use of following formulas:

$$\left.\frac{d\bar{V}}{dt}\right|_{t=t_i} \approx \frac{1}{2}(\bar{V}_{i+1} - \bar{V}_{i-1}) \tag{8.2}$$

$$\left.\frac{d^2\bar{V}}{dt^2}\right|_{t=t_i} \approx \bar{V}_{i+1} - 2\bar{V}_i + \bar{V}_{i-1} \tag{8.3}$$

$$\left.\frac{dw}{dt}\right|_{t=t_i} \approx \frac{1}{14}(w_{i+1} - w_{i-1}) \tag{8.4}$$

$$\left.\frac{d^2w}{dt^2}\right|_{t=t_i} \approx \frac{1}{49}(w_{i+1} - 2w_i + w_{i-1}) \tag{8.5}$$

Coefficients 1/14 and 1/49 have appeared in formulas (8.4) and (8.5) due to the fact that we will always measure time in days. For the first epidemic waves, Eqs. (8.2) and (8.4) yield estimations of dV/dt (see (4.11) and (4.12)); Eqs. (8.3) and (8.5) allow estimating d^2V/dt^2 (see (4.26) and (4.27)). In particular, second derivatives (8.3) and (8.5) have to change their signs from positive to negative at the time moment t_s, which can be calculated with the use of (4.31). Thus, the minimums of distributions (8.2) and (8.4) and zero points of (8.3) and (8.5) can be treated as moments when the parameters of epidemic change. We will analyze the distributions (8.2)–(8.5) for some countries and global statistics as well.

As usual, we will use the WHO official accumulated number of cases. Some data sets are shown in many tables from previous chapters. In addition, some recent data sets for daily (V_j) and weekly (w_j) accumulated number of cases in Ukraine and in

Table 8.1 Official cumulative numbers of confirmed COVID-19 cases in Ukraine and the world, [1, 93]

Day in July and August 2020	Time in days t_j	Number of cases in Ukraine V_j	Global number of cases V_j	Day in August, September and October 2020	Time in days t_j	Number of cases in Ukraine w_j	Global number of cases w_j
30	192	69,884	17,106,007	22	215	104,958	23,057,288
31	193	71,056	17,396,943	29	222	119,074	24,854,140
1	194	72,168	17,660,523	5	229	135,894	26,763,217
2	195	73,158	17,918,582	12	236	154,335	28,637,952
3	196	74,219	18,142,718	19	243	175,678	30,675,675
4	197	75,490	18,354,342	26	250	198,634	32,730,945
5	198	76,808	18,614,177	3	257	226,462	34,804,348
6	199	78,261	18,902,735	10	264	261,035	37,109,851
7	200	79,750	19,187,943				
8	201	80,949	19,462,112				
9	202	81,957	19,718,030				
10	203	83,115	19,936,210				
11	204	84,548	20,162,474				
12	205	86,140	20,439,814				
13	206	87,872	20,730,456				
14	207	89,719	21,026,758				
15	208	91,356	21,294,845				

the world and corresponding moments of time t_j (in days starting with January 20, 2020) are shown in Table 8.1. Figures 8.1, 8.2, 8.3, 8.4, 8.5, 8.6, 8.7, 8.8, 8.9, 8.10, 8.11, 8.12 and 8.13 present the results of calculations.

Figure 8.1 illustrates that the global pandemic dynamics was more or less similar to SIR one in the fist half of April. The jumps of the second derivative ("stars") are already visible in the second half of April, and the most intense maximums of this characteristic correspond to May 12 and 28; June 13; July 5, 14 and 29; the beginning of October 2020. The first derivative ("triangles" in Fig. 8.1) increased almost monotonously. Some visible minimum of dV/dt was achieved only in July (the same picture we can see also in Fig. 7.6), but the corresponding values are very high and not much lower than the maximum level in late July. Therefore, two predictions for the first wave of global pandemic dynamics presented in Table 6.25 cannot yield satisfactory accuracy as evidenced by high levels of relative error $\varepsilon > 80\%$. The time moments corresponding the maximum values of second derivatives (8.3) or (8.5) can be treated as the starting points of the second, third and so on waves of the pandemic. We will try to simulate these waves in the next sections.

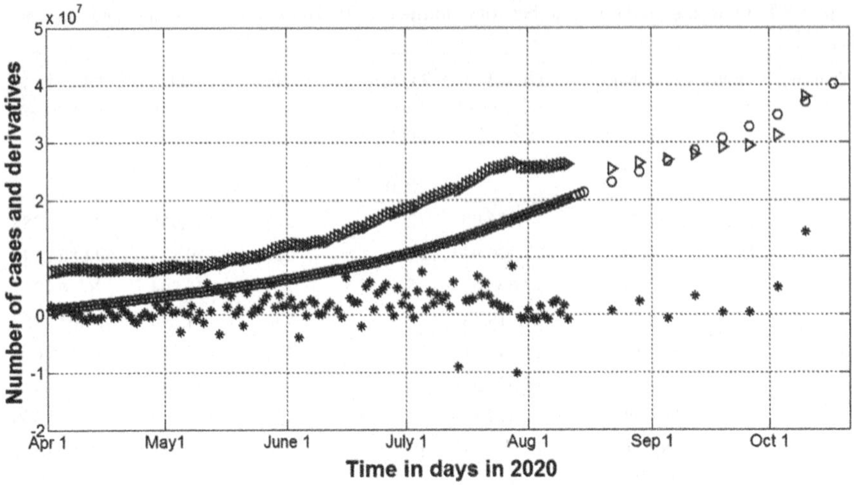

Fig. 8.1 Pandemic dynamics in the world versus time in days. Accumulated number of cases (V_j and w_j—"circles"; smoothed values—line, Eq. (8.1)). "Triangles" show the first derivative (Eqs. (8.2) and (8.4)) multiplied by 100, "stars"—the second derivative (Eqs. (8.3) and (8.5)) multiplied by 1000

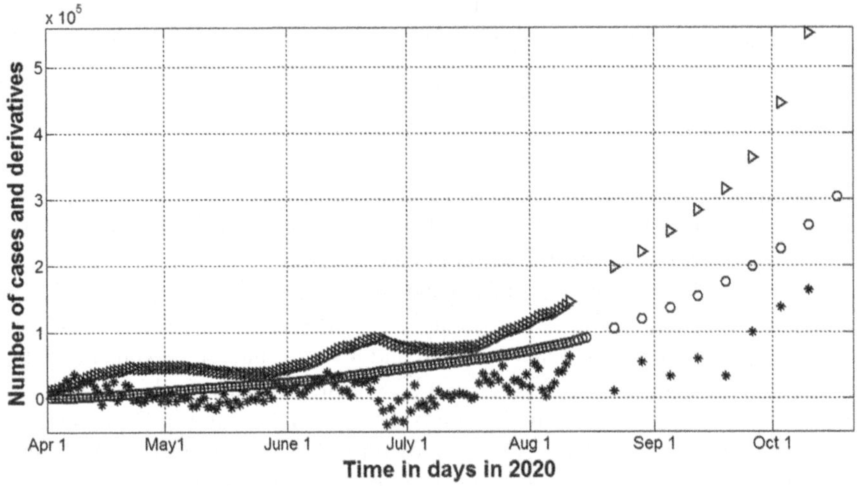

Fig. 8.2 Pandemic dynamics in Ukraine versus time in days. Accumulated number of cases (V_j and w_j—"circles"; smoothed values—line, Eq. (8.1)). "Triangles" show the first derivative (Eqs. (8.2) and (8.4)) multiplied by 100, "stars"—the second derivative (Eqs. (8.3) and (8.5)) multiplied by 1000

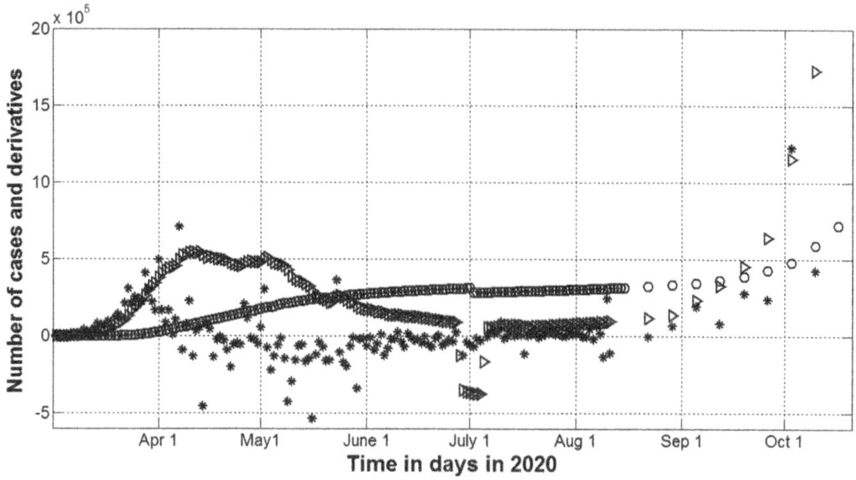

Fig. 8.3 Pandemic dynamics in the UK versus time in days. Accumulated number of cases (V_j and w_j—"circles"; smoothed values—line, Eq. (8.1)). "Triangles" show the first derivative (Eqs. (8.2) and (8.4)) multiplied by 100, "stars"—the second derivative (Eqs. (8.3) and (8.5)) multiplied by 1000

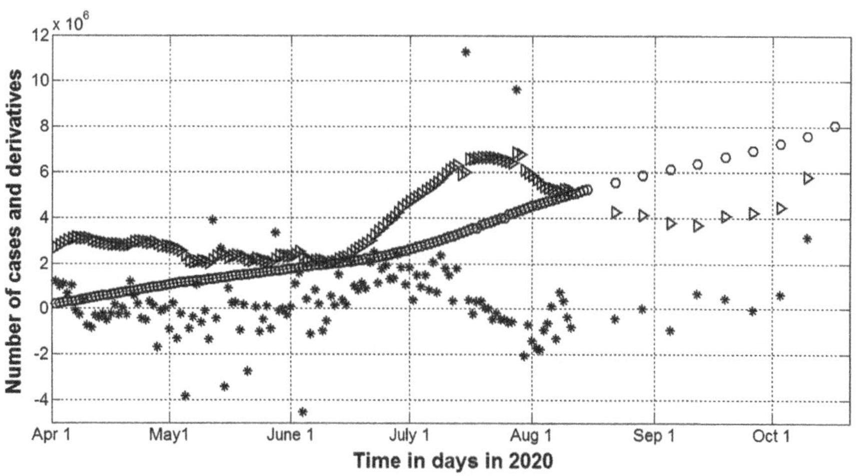

Fig. 8.4 Pandemic dynamics in USA versus time in days. Accumulated number of cases (V_j and w_j—"circles"; smoothed values—line, Eq. (8.1)). "Triangles" show the first derivative (Eqs. (8.2) and (8.4)) multiplied by 100, "stars"—the second derivative (Eqs. (8.3) and (8.5)) multiplied by 1000

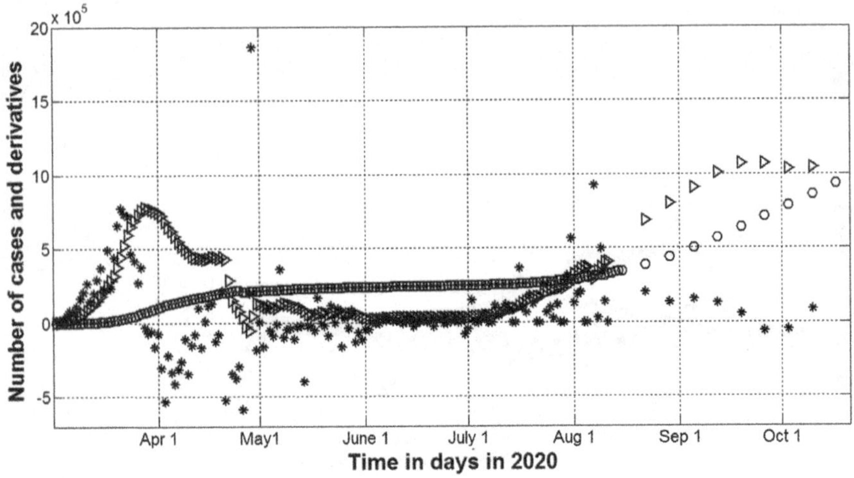

Fig. 8.5 Pandemic dynamics in Spain versus time in days. Accumulated number of cases (V_j and w_j—"circles"; smoothed values—line, Eq. (8.1)). "Triangles" show the first derivative (Eqs. (8.2) and (8.4)) multiplied by 100, "stars"—the second derivative (Eqs. (8.3) and (8.5)) multiplied by 1000

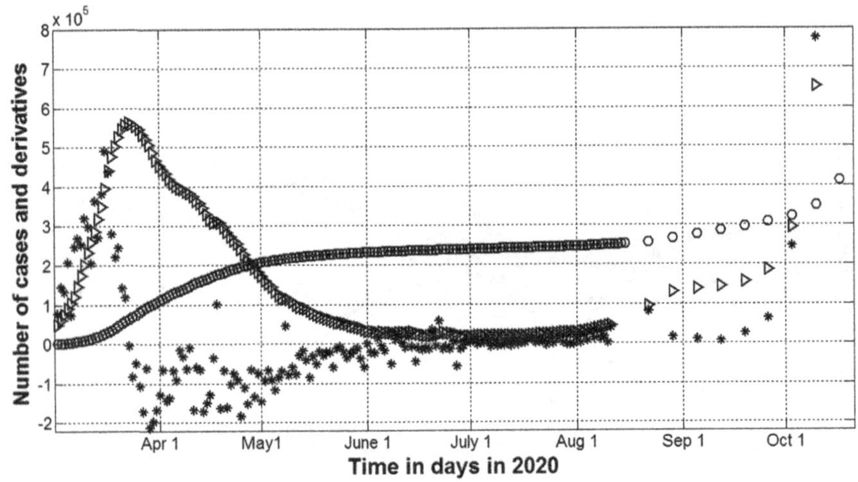

Fig. 8.6 Pandemic dynamics in Italy versus time in days. Accumulated number of cases (V_j and w_j—"circles"; smoothed values—line, Eq. (8.1)). "Triangles" show the first derivative (Eqs. (8.2) and (8.4)) multiplied by 100, "stars"—the second derivative (Eqs. (8.3) and (8.5)) multiplied by 1000

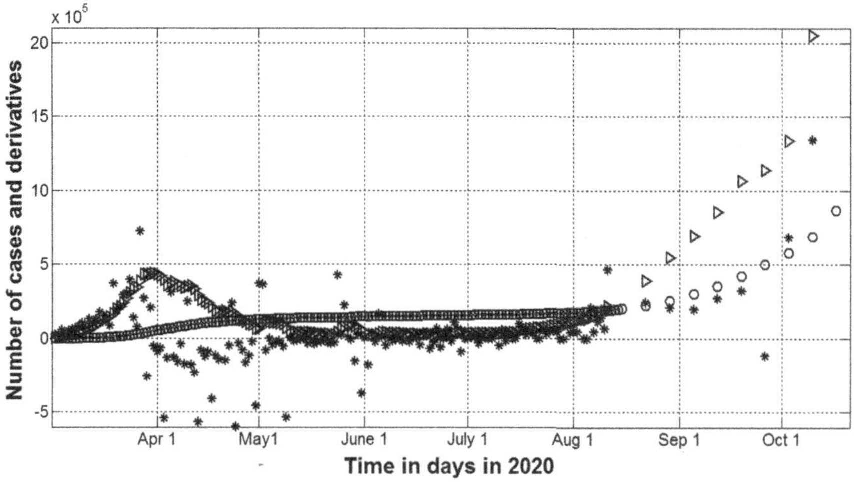

Fig. 8.7 Pandemic dynamics in France versus time in days. Accumulated number of cases (V_j and w_j—"circles"; smoothed values—line, Eq. (8.1)). "Triangles" show the first derivative (Eqs. (8.2) and (8.4)) multiplied by 100, "stars"—the second derivative (Eqs. (8.3) and (8.5)) multiplied by 1000

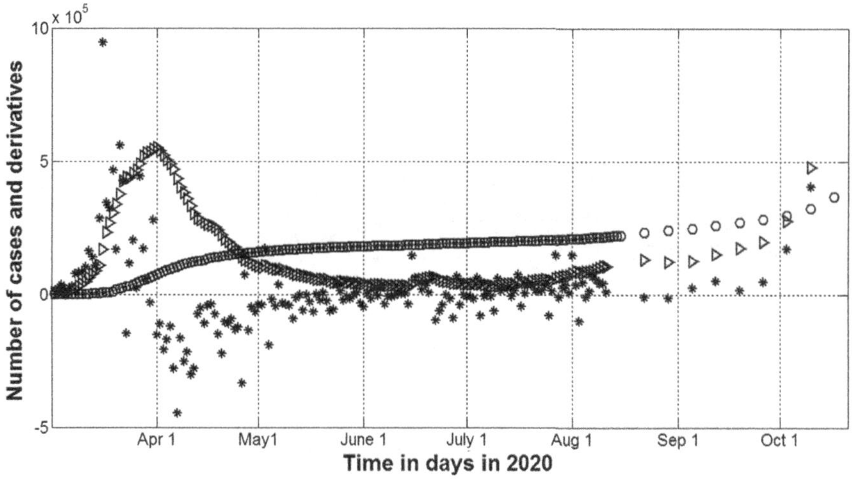

Fig. 8.8 Pandemic dynamics in Germany versus time in days. Accumulated number of cases (V_j and w_j—"circles"; smoothed values—line, Eq. (8.1)). "Triangles" show the first derivative (Eqs. (8.2) and (8.4)) multiplied by 100, "stars"—the second derivative (Eqs. (8.3) and (8.5)) multiplied by 1000

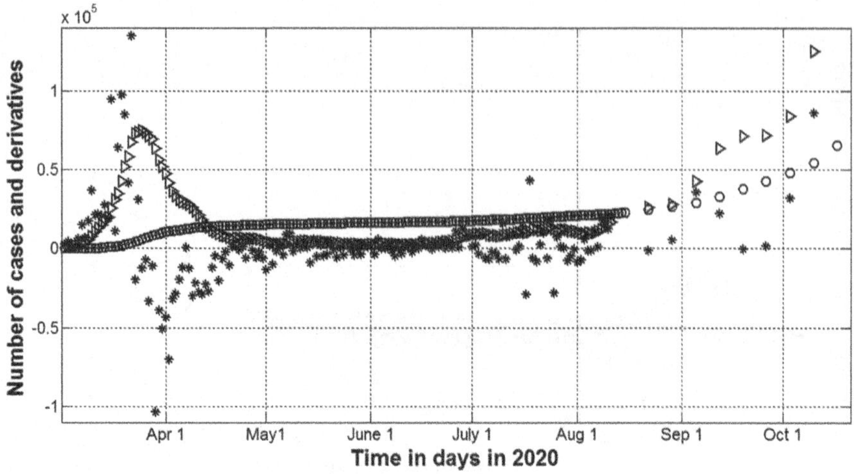

Fig. 8.9 Pandemic dynamics in Austria versus time in days. Accumulated number of cases (V_j and w_j—"circles"; smoothed values—line, Eq. (8.1)). "Triangles" show the first derivative (Eqs. (8.2) and (8.4)) multiplied by 100, "stars"—the second derivative (Eqs. (8.3) and (8.5)) multiplied by 1000

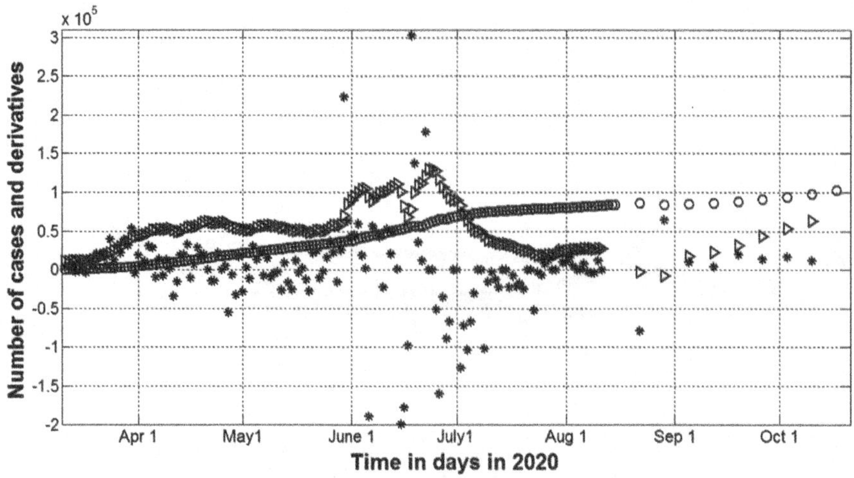

Fig. 8.10 Pandemic dynamics in Sweden versus time in days. Accumulated number of cases (V_j and w_j—"circles"; smoothed values—line, Eq. (8.1)). "Triangles" show the first derivative (Eqs. (8.2) and (8.4)) multiplied by 100, "stars"—the second derivative (Eqs. (8.3) and (8.5)) multiplied by 1000

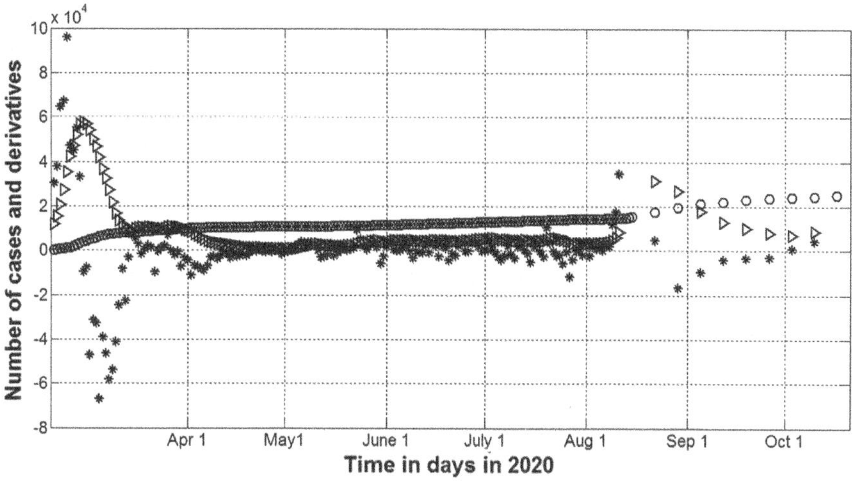

Fig. 8.11 Pandemic dynamics in the Republic of Korea versus time in days. Accumulated number of cases (V_j and w_j—"circles"; smoothed values—line, Eq. (8.1)). "Triangles" show the first derivative (Eqs. (8.2) and (8.4)) multiplied by 100, "stars"—the second derivative (Eqs. (8.3) and (8.5)) multiplied by 1000

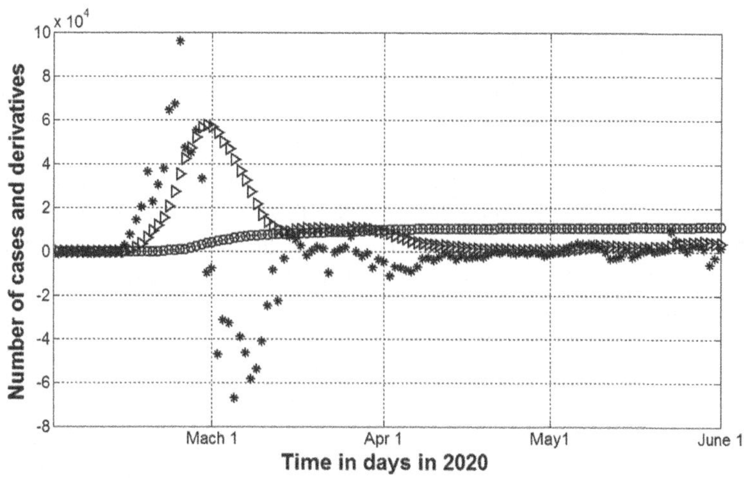

Fig. 8.12 Pandemic dynamics in the Republic of Korea before June 1, 2020, versus time in days. Accumulated number of cases (V_j and w_j—"circles"; smoothed values—line, Eq. (8.1)). "Triangles" show the first derivative (Eqs. (8.2) and (8.4)) multiplied by 100, "stars"—the second derivative (Eqs. (8.3) and (8.5)) multiplied by 1000

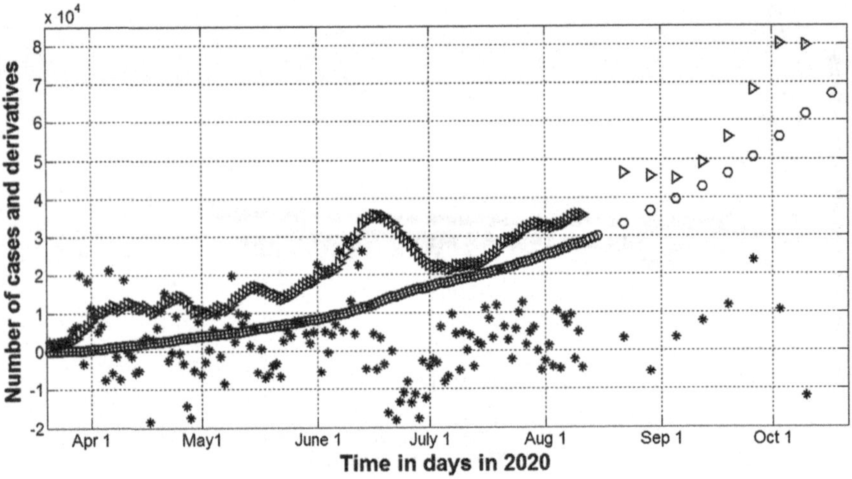

Fig. 8.13 Pandemic dynamics in the Republic of Moldova versus time in days. Accumulated number of cases (V_j and w_j—"circles"; smoothed values—line, Eq. (8.1)). "Triangles" show the first derivative (Eqs. (8.2) and (8.4)) multiplied by 100, "stars"—the second derivative (Eqs. (8.3) and (8.5)) multiplied by 1000

Global dynamics is influenced by many factors, which can make it very irregular. The development of the epidemic in some countries may be more similar to the picture described by SIR model. But the corresponding picture can be obtained only if the registered number of cases is close to the real one. In countries where the problems with testing (and, accordingly, with the isolation of patients) occurred, one should not expect good simulation results. Unfortunately, Ukraine is one of such countries, so even the eighth forecast (see Table 6.23) made using data for the period T_c immediately preceding the first explicit maximum of the second derivative on May 16 (visible in Fig. 8.2) did not provide good accuracy (ε = 89%).

We can threat the days: May 16, May 29, June 8, July 3 and 19, August 2 and 18; the end of August and September as the beginning of the second, third, etc., waves of the COVID-19 pandemic in Ukraine (see Fig. 8.2). Some of these days are located very close, which reduces the number of observations of the corresponding waves and the accuracy of calculations. So in papers [89, 94], some of these waves were combined. The presence of epidemic waves in Ukraine is also evidenced by two clear minimums of dV/dt (see "triangles in Fig. 8.2), but the corresponding values are very high and do not differ very much from the maximum daily number of new cases typical for early May and mid-June. We will focus on the analysis of the next epidemic waves in Ukraine in Chap. 11.

The proposed method is very sensitive to data irregularities and can be also used for their detection. For example, the number of accumulated cases in the UK (reported to WHO on May 20) was lower than on May 19. The same situation occurred between July 1 and 2. The corresponding irregularities in the values of the first and second derivatives are visible in Fig. 8.3. Neglecting them, we can conclude that the beginning of the second wave of the epidemic in the UK occurred in late April, but the corresponding values of the first derivative d*V*/d*t* ("triangles" in Fig. 8.3) were quite high, which reduced the accuracy of SIR simulation. The corresponding value of relative error $\varepsilon = 33.4\%$ (see Table 6.16) is lower than for Ukraine, but much higher than for the EU countries. The next epidemic waves in the UK probably started in early August, September, and October.

The number of accumulated cases in USA (reported to WHO on May 8) was lower than on May 9. The same incorrectness occurred between May 18 and 19. For dates: July 17–18 and July 24–25, the numbers of cases was the same (no increase). All these irregularities are visible in Fig. 8.4 (in particular, there are very high values of the second derivative ("stars"). Neglecting them, we can conclude that the beginning of the second epidemic wave in USA occurred on April 21, with the very high values of the first derivative d*V*/d*t* (in comparison with the highest level in early April, see "triangles" in Fig. 8.4). The period taken for SIR simulations (T_c: April 9–29, 2020, see Table 6.17) covers days from both the first and second waves.

These facts have reduced the accuracy of the simulation ($\varepsilon = 78.2\%$). The final day of the first pandemic wave in USA was estimated to occur on September 23, 2020 (later than in the EU countries, see Table 7.6). Between April 21 and this date, at least six changes in epidemic dynamics are visible in Fig. 8.4. This fact also reduced the accuracy of prediction presented in Table 6.17 which is valid for the first wave only. Unfortunately, a new severe wave has started in USA in mid-October (see last "star" in Fig. 8.4).

The number of accumulated cases in Spain (reported to WHO) was very irregular in periods April 24–May 5; May 24–25; July 3–5: July 10–12; July 31–August 2 and August 7–10 (there were no monotonic increase). These irregularities are visible in Fig. 8.5 (in particular, there are some negative values of the first derivative ("triangles")). Neglecting them, we can conclude that the beginning of a weak second wave of the epidemic in Spain occurred in mid-April, with not very high values of the first derivative d*V*/d*t* (in comparison with the highest level in late March, see "triangles" in Fig. 8.5). In June, the values of the first derivative were much closer to zero which testifies to the successful overcoming of the first severe epidemic wave in Spain. Nevertheless in July, the first derivative (and the daily increase in the number of cases) began to increase very fast, and Spain was the first EU country to cross the 1 million registered cases.

We have used the period T_c: March 29–April 18 (which corresponds to the first wave), to calculate the second prediction for Spain and have obtained very high accuracy $\varepsilon = 12.2\%$ (see Table 6.13). The predicted value of t_{final}(corresponding to July 7) agrees with the nature of the epidemic dynamics, which is characterized by very low values of the new daily number of cases (first derivative d*V*/d*t*) until early July 2020 (see "triangles in Fig. 8.5). As of July 7, the average value of d*V*/d*t* can be estimated with the use of accumulated numbers of cases 249271 (June 30) and 255953 (July 13). The result is 514 new cases per day. According to (4.11) and (4.17):

$$\frac{dV}{dt} = \alpha SI \rightarrow \alpha S_{\infty} \quad \text{when} \quad t \rightarrow t_{\text{final}} \tag{8.6}$$

Putting in (8.6) the corresponding values from Table 6.13 (prediction 2), we can see that the theoretical estimation $dV/dt \rightarrow 0.576$ drastically differs from the real number of new cases. This fact testifies that quarantine weakening and changes in social behavior after the period T_c caused many additional COVID-19 cases and new epidemic waves.

In comparison with Spain, Italy reported some increase in the number of cases every day. Thus, we can treat three jumps of second derivatives in April, May and June (see "stars" in Fig. 8.6) as the beginnings of some weak new epidemic waves caused by changes in quarantine restrictions and social behavior. The average daily number of new cases decreased almost monotonously form Mach 21 till mid-June, but started to increase monotonously after July 12 (see "triangles" in Fig. 8.6). The predicted value of t_{final} corresponds to August 18 (see Table 6.7, prediction 6), i.e., the next severe wave of epidemic in Italy has started before the predicted first wave has finished. After August 2020, the daily rate dV/dt and second derivative became very large. We have used the period T_c: March 29–April 18 (which corresponds to the first wave before jump in the second derivative), to calculate prediction 6 and have obtained very high accuracy $\varepsilon = 11.6\%$ (see Table 6.7). The relative errors of previous predictions (with earlier periods T_c) were higher (see Table 6.4), because many cases were not detected.

As of August 18, the average value of dV/dt can be estimated with the use of accumulated numbers of cases 250,103 (August 8) and 258,136 (August 22). The result is 617.9 new cases per day. According to (8.6) and Table 6.7 (prediction 6), the theory yields $dV/dt \rightarrow 0.608$. As in the case of Spain, the theoretical estimation for Italy is much lower than the real number of new cases (in spite of the good accuracy of predictions of the final size of the epidemics in both countries). This fact testifies that quarantine weakening and changes in social behavior after the period T_c caused many additional COVID-19 cases and can be the reason of the new severe wave/waves in autumn-winter 2020–2021.

There are some irregularities in the reported number of accumulated cases in France (no monotonic increase in periods April 27–29, June 1–3, June 23–25, and August 7–9), which are visible in Fig. 8.7. Neglecting them, we can conclude that some changes in epidemic dynamics occurred already in early April. This conclusion is supported by jumps in the values of the second derivative ("stars") and weak increase of the first derivative dV/dt (see "triangles" in Fig. 8.7). Nevertheless, the average daily numbers of new cases were rather low in June and early July in comparison with the highest level in late March.

Let us compare these figures with the theoretical estimations presented in Table 6.14. The highest accuracy $\varepsilon = 16\%$ corresponds to prediction 2 calculated with the use of T_c: April 5–18, 2020. During this period, some changes in the epidemic dynamics already occurred. Probably, it is the reason of the lower accuracy in comparison with predictions for Italy and Spain. Since June 20, 2020, was predicted as the last day of the first epidemic wave in France (see Table 6.14), the corresponding average value of dV/dt can be estimated with the use of accumulated numbers of cases 153356 (June 17) and 155256 (June 23). The result—316.7 new cases per day—is lower than in Spain and Italy, but the theory yields much lower figure. According to (8.6) and Table 6.14(prediction 2), $dV/dt \to 1.1$. As in the cases of Spain and Italy, the much higher real number of new cases testifies that quarantine weakening and changes in social behavior after the period T_c caused many additional COVID-19 cases and can be the reason of a new severe wave in September–October 2020, visible in Fig. 8.7.

In German reports for WHO, there were six pairs of days with the same accumulated number of cases (April 29–30; June 17–18; June 30–July 1; July 12–13, 25–26, 28–29). These peculiarities cased some high values of the second derivative (see "stars" in Fig. 8.8). Neglecting them, we can see a monotonic decrease in the daily number of cases in April, May and early June (see "triangles" in Fig. 8.8). It means that during this rather long period, there were no visible changes in epidemic dynamics. But situation changed after mid-June probably due to quarantine weakening and changes in social behavior.

We have used the period T_c: April 9–29, to calculate the second prediction for Germany and have obtained the value of relative accuracy $\varepsilon = 16.5\%$ (see Table 6.15). The predicted value of t_{final}(August 4) corresponds to the period with increasing average daily numbers of new cases. It means that next epidemic wave/waves started in Germany before the finishing the first one. As of August 4, the average value of dV/dt can be estimated with the use of accumulated numbers of cases 206 926 (July 28) and 218 519 (August 11). The result is 828 new cases per day is much higher than in Spain, Italy and France. According to (8.6) and Table 6.15 (prediction 2), $dV/dt \to 0.661$. As in the cases of other EU countries, the much higher real number of new cases testifies that quarantine weakening and changes in social behavior after the period T_c caused many additional COVID-19 cases and can be the reason of a new severe wave in October 2020, visible in Fig. 8.8.

The number of cases in Austria corresponding May 5 is lower than for the previous day. There is also the pair of days with the same accumulated number of cases (July 20–21). These peculiarities cased some high values of the second derivative (see "stars" in Fig. 8.9). It is visible the change in dV/dt slope which occurred on April 23 (see "triangles" in Fig. 8.9). Probably, this can be explained by the fact that small shops were opened in Austria after April 14, 2020. After some incubation period, the epidemic dynamics has changed. In May and June, the average daily numbers of new cases were rather low in comparison with the highest level in late March. But situation changed in late June (probably due to quarantine weakening and changes in social behavior).

We have used the period T_c: March 28–April 10 to calculate prediction 6 and have obtained the very high value of relative accuracy $\varepsilon = 6.6\%$ (see Table 6.12). The previous predictions (presented in Sect. 6.4) yielded higher errors. Probably, difference between the numbers of registered cases and real ones was larger before this period. The predicted value of t_{final}(May 18) corresponds to the period with decreasing values of dV/dt. It means that next epidemic waves (in July, August, September and October; visible in Fig. 8.9) started in Austria after finishing the first one. As of May 18, the average value of dV/dt can be estimated with the use of accumulated numbers of cases 15874 (May 11) and 16459 (May 25). The result—41.8 new cases per day—is the lowest among all considered EU countries. According to (8.6) and Table 6.12 (prediction 6), $dV/dt \to 1.156$. As in other EU countries, the much higher real number of new cases testifies that quarantine weakening and changes in social behavior after the period T_c caused many additional COVID-19 cases and new epidemic waves in Austria.

It is very interesting to compare the dynamics of the pandemic in Sweden, where no total lockdown was introduced (in comparison with other European countries, where it was used and quarantine restrictions were much severe). There are some irregularities in the reported number of accumulated cases in Sweden. There were no increase in V_j figures in periods (June 18–21, 26–28; July 3–5, 9–12, 17–19, 24–26; August 1–2, 7–9, and 14–15). The w_i value corresponding to August 29 is less than to August 22. These irregularities decrease the accuracy of calculations with the use of formulas (8.2)–(8.5). But even before mid-June, the epidemic dynamics significantly differs from the situation in other EU countries, the UK and Ukraine (compare Fig. 8.10 with Figs. 8.2, 8.3, 8.5, 8.6, 8.7 and 8.9). For example, the values of the second derivative ("stars") are very random in the case of Sweden. In the above-mentioned countries, they were mostly positive after the epidemic outbreak and then became mostly negative. A similar nature of the second derivatives is characteristic of USA (see Fig. 8.4).

The minimum of the average daily number of new cases (dV/dt = 188) was achieved in Sweden only on July 24, 2020 (later than in other EU countries, compare "triangles" in Figs. 8.5, 8.6, 8.7, 8.8, 8.9 and 8.10). The increase of this number in September and October in Sweden was not as rapid as in other EU countries (compare "triangles" in Figs. 8.5, 8.6, 8.7, 8.8, 8.9 and 8.10). To estimate the efficiency of Swedish strategy, we can compare the number of accumulated cases per one million of population. As of September 26, 2020, this figure in Sweden was 9,003 [86]. Comparison with the values presented in Table 7.4 shows that higher levels were only in the USA, Moldova and Spain. But as of October 17, 2020, higher numbers of cases per capita were also in France (13,298) and the UK (10,642) (in comparison with 10,219 in Sweden). It must be noted that some other EU (countries Belgium, the Netherlands, Czechia, Luxemburg) had higher levels of cases per capita, and the levels in Montenegro, Andorra and San Marino were more than twice higher. Therefore, final conclusions about the success of the Swedish strategy can be made only after the COVID-19 pandemic completion.

We did not notice any evident inaccuracies in the data provided by South Korea to the WHO (unlike some of the other countries analyzed above). Therefore, fluctuations in the values of the first and second derivative (see "triangles" and "stars" in Fig. 8.11) can be interpreted as consequences of changes in the epidemic dynamics. In particular, the character of pandemic dynamics changed in mid-March; in August, we can see rather fast increase in the daily number of cases (see "triangles" in Fig. 8.11).

We have used the two different periods T_c: February 17–March 12 and April 9–29 to calculate two predictions for South Korea and have obtained the rather high values of relative accuracy $\varepsilon = 7.8\%$ and 15.3% but very different optimal values of some SIR parameters (see Table 6.2). The first period corresponds to the initial stage of the epidemic when many cases were not detected. This fact usually causes the limited accuracy of predictions. The second T_c corresponds to the period after many changes in epidemic dynamics (visible in Fig. 8.12). Therefore, the theory used in Chap. 6 (applicable only for the first waves of epidemic) cannot yield good accuracy. That is why very different values of the optimal parameters were obtained for two predictions (see Table 6.2).

If we select the first prediction, the value of t_{final} (March 20, 2020) corresponds to the period with a local minimum of $\mathrm{d}V/\mathrm{d}t$ (around 90 new cases per day, see "triangles" in Fig. 8.11). So we can conclude that the first epidemic wave in South Korea has finished in the mid-March and was controllable during rather a long period of time. It looks that the sharp increase in the number of new cases (which occurred in August, see "triangles" in Fig. 8.11) was also overcome.

In the reports of the Republic of Moldova, there were only two pairs of days with the same accumulated number of cases (April 2–3 and July 17–18). Nevertheless, the values of the first and second derivatives are very irregular (see "triangles" and "stars" in Fig. 8.13). We saw a similar picture in the case of Sweden (Fig. 8.10), but in Moldova the national lockdown was used. Probably, numerous quarantine violations, its early easing and testing problems led to a situation where the first wave of the epidemic did not end, and its new waves led to an almost constant increase in the number of new cases.

We have used the two different periods T_c: March 28–April 10 and April 5–18 to calculate two predictions for Moldova and have obtained very low accuracy $\varepsilon = 77$ and 71%. Probably, both periods correspond to the initial stage of the epidemic when many cases were not detected. The values of t_{final} (June 11 and 16) correspond to the period with the absolute maximum of $\mathrm{d}V/\mathrm{d}t$ (around 90 new cases per day, see "triangles" in Fig. 8.13). So we can conclude that both SIR simulations were not successful in the case Moldova. Probably, it will be possible to improve the accuracy with the use of generalized SIR theory (see next Chapters), which allows simulating different epidemic waves.

The idea of comparing the epidemic dynamics in the country and its individual regions presented in [90] looks very interesting. Detailed research on this issue is still ahead. Here we present the results of the comparison for Ukraine and its capital Kyiv. To do this, we use data on the accumulated number of cases from the previous tables and some recent data from Table 8.2.

Table 8.2 Official cumulative numbers of confirmed COVID-19 cases in Ukraine and Kyiv, [95, 96]

Day in August and September 2020	Time in days t_j	Number of cases in Ukraine V_j	Number of cases Kyiv V_j	Day in September and October 2020	Time in days t_j	Number of cases in Ukraine V_j	Number of cases Kyiv V_j
16	209	92,820	10,642	17	241	169,472	18,740
17	210	94,436	10,730	18	242	172,712	19,140
18	211	96,403	10,945	19	243	175,678	19,483
19	212	98,537	11,172	20	244	178,353	19,758
20	213	100,643	11,328	21	245	181,237	20,028
21	214	102,971	11,507	22	246	184,734	20,335
22	215	104,958	11,814	23	247	188,106	20,665
23	216	106,757	12,023	24	248	191,671	21,045
24	217	108,415	12,153	25	249	195,504	21,492
25	218	110,085	12,248	26	250	198,634	21,815
26	219	112,059	12,468	27	251	201,305	22,143
27	220	114,497	12,719	28	252	204,932	22,559
28	221	116,978	12,984	29	253	208,959	22,778
29	222	119,074	13,093	30	254	213,028	23,155
30	223	121,215	13,370	1	255	217,661	23,602
31	224	123,303	13,546	2	256	222,322	24,043
1	225	125,798	13,878	3	257	226,462	24,450
2	226	128,228	14,158	4	258	230,236	24,736
3	227	130,951	14,477	5	259	234,584	25,091
4	228	133,787	14,792	6	260	239,337	25,474
5	229	135,894	15,032	7	261	244,734	25,896
6	230	138,068	15,276	8	262	250,538	26,463
7	231	140,479	15,511	9	263	256,266	26,995
8	232	143,030	15,821	10	264	261,034	27,497
9	233	145,612	16,119	11	265	265,454	27,693
10	234	148,756	16,523	12	266	270,587	28,090
11	235	151,859	16,951	13	267	276,177	28,512
12	236	154,335	17,261	14	268	281,239	28,957
13	237	156,797	17,429	15	269	287,231	29,412
14	238	159,702	17,714	16	270	293,641	30,010
15	239	162,660	18,070	17	271	298,872	30,463
16	240	166,244	18,449	18	272	303,638	30,848

Figures 8.14 and 8.15 represent the results of calculations for Ukraine (in red) and Kyiv (in blue) with the use of Eq. (8.1) (lines); Eqs. (8.2) and (8.3)—small "triangles" and "stars". All these calculations use the daily number of accumulated cases V_j. The use of weekly number of cases w_i (Eqs. (8.4) and (8.5)) is shown in Fig. 8.14 by large markers. It can be seen that w_i can be successively used to

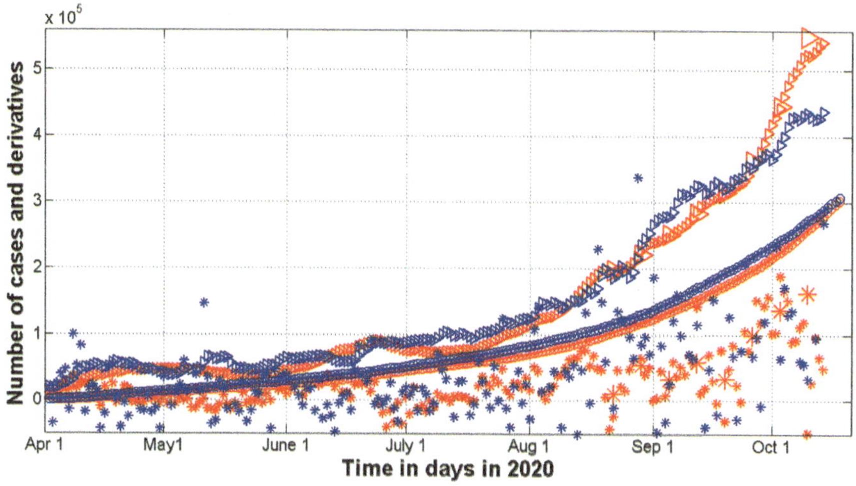

Fig. 8.14 Comparison of the pandemic dynamics in Ukraine (red) and Kyiv (blue) versus time in days. Accumulated number of cases (V_j—"circles"; smoothed values—lines, Eq. (8.1); for Kyiv multiplied by 10). "Triangles" show the first derivative calculated with the use of Eqs. (8.2) and (8.4) (only for Ukraine, large markers) multiplied by 100 for Ukraine and by 1000 for Kyiv. "Stars" represent the second derivative calculated with the use of Eqs. (8.3) and (8.5) (only for Ukraine, large markers) multiplied by 1000 for Ukraine and by 10,000 for Kyiv

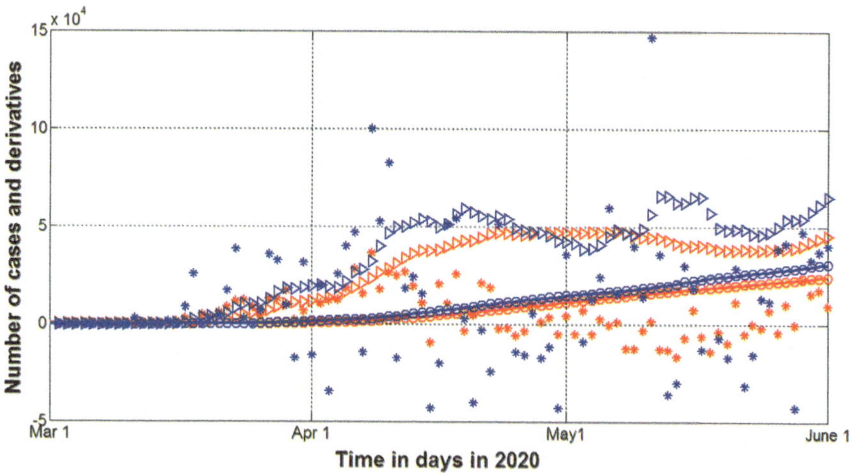

Fig. 8.15 Comparison of the pandemic dynamics in Ukraine (red) and Kyiv (blue) before June 1, 2020. Accumulated number of cases (V_j—"circles"; smoothed values—lines, Eq. (8.1); for Kyiv multiplied by 10). "Triangles" show the first derivative calculated with the use of Eq. (8.2) multiplied by 100 for Ukraine and by 1000 for Kyiv. "Stars" represent the second derivative calculated with the use of Eq. (8.3) multiplied by 1000 for Ukraine and by 10,000 for Kyiv

calculate the first derivative dV/dt, but for the second derivative it is better to use the V_j values.

Figures 8.14 and 8.15 show that after mid-April the number of cases in the capital exceeded 10% of the number in Ukraine. The population of Kyiv is about 6.8% of the population of Ukraine. The higher number of cases in the capital can be explained by greater mobility of the population, higher amount of guests and may be better situation with testing. In many countries, the largest cities became the epicenters of the pandemic.

The dynamics of the epidemic in Kyiv is much more chaotic than in Ukraine as a whole. This is evidenced by the relatively large values of the second derivative (compare "stars" in Figs. 8.14 and 8.15) and the nature of the dependences for the first derivative (compare "triangles" in Figs. 8.14 and 8.15). This is probably due to the random nature of V_j values. As the population of Ukraine (and the number of registered cases) is much larger, random factors were less noticeable. Nevertheless, the case of Moldova shown in Fig. 8.13 looks more chaotic, despite the larger population than in Kyiv.

Figure 8.15 illustrates that the average daily number of new cases in Kyiv significantly increased in mid-May. This is probably due to the cancelation of the lockdown on May 10, 2020 and mass violations of quarantine in early May. In Ukraine as a whole, this trend in dV/dt values is almost invisible, but we can see a jump in the second derivative values (see red "stars" in Fig. 8.15). The first minimum of the first derivative dV/dt in Ukraine coincided with the second minimum for Kyiv. We can see that the second epidemic wave in Kyiv started in early May before the first wave was finished.

We have used three different periods T_c: March 28–April 10; April 11–24 and May 13–26 to calculate predictions for Kyiv and have obtained very low accuracy ε = 57–73% (see Table 6.24). Two first predictions use the periods T_c corresponding the first wave of the epidemic in Kyiv. But there were problems and changes in testing algorithm in April; therefore, the numbers of registered cases were far from the real one. Prediction 3 was calculated with the use of data set corresponding to the second epidemic wave; therefore, the theory applied in Chap. 6 cannot yield good accuracy. The error of this prediction is the highest. So we can conclude that SIR simulations were not successful in the case Kyiv, but they showed that testing improving causes the drastic increase in the estimations of the final size of the epidemic. Probably, it will be possible to improve the accuracy with the use of generalized SIR theory (see next Chapters), which allows simulating of different epidemic waves.

The results of this chapter show that smoothing dependence of the accumulated number of cases $V_j(t_j)$ and its differentiation (Eqs. (8.1)–(8.5)) can provide fairly accurate and useful information about the course of the epidemic, identify important changes in its dynamics and provide timely recommendations for quarantine measures or control of social distancing.

Chapter 9
General SIR Model and Its Exact Solution

In order to simulate different epidemic waves, we will propose a generalized SIR model which assumes constant value of its parameters for every separate wave and present an analytical solution of corresponding equations.

In previous chapters, we have demonstrated that dynamics of COVID-19 pandemic can be very different. Changes in quarantine conditions, human behavior, pathogen activity, weather, etc., can cause changes in the course of the pandemic, namely the so-called epidemic waves. The mathematical simulation of these waves needs development of some criteria of the wave detection presented in previous chapter. Here, we will modify the SIR model to make it applicable for simulations of different epidemic waves. Some results have been already published in [89, 94].

We suppose that the SIR model parameters are constant for every epidemic wave, i.e., for the time periods: $t_i^* \le t \le t_{i+1}^*,\ i = 1, 2, 3, \ldots$. Then for every wave, we can use the equations, similar to (4.1)–(4.3):

$$\frac{dS}{dt} = -\alpha_i SI, \tag{9.1}$$

$$\frac{dI}{dt} = \alpha_i SI - \rho_i I, \tag{9.2}$$

$$\frac{dR}{dt} = \rho_i I. \tag{9.3}$$

As before, S is the number of susceptible persons (who are sensitive to the pathogen and **not protected**), I is the number of infected persons (who are sick and **spread the infection**; please do not confuse with the number of still ill persons, so known active cases) and R is the number of removed persons (who **no longer spread the infection**; this number is the sum of isolated, recovered, dead and infected people who left the region). Parameters α_i and ρ_i are supposed to be constant for every epidemic wave.

I. Nesteruk, *COVID-19 Pandemic Dynamics*,
https://doi.org/10.1007/978-981-33-6416-5_9

As mentioned in the chap. 4, parameters α_i show how quick the susceptible persons become infected (see (9.1)). Large values of this parameter correspond to severe epidemics with many victims. These parameters accumulate many characteristics. First, they show how strong (virulent) is the pathogen and what is the way of its spreading. Parameters α_i accumulate also the frequency of contacts and the way of contacting. In order to decrease the values of α_i, we have to minimize the number of our contacts and change our contacting habits. For example, we have to avoid the public places, use masks there, and minimize or cancel traveling. We have to change our contact habits: to avoid handshakes and kisses. First, all these simple things are very useful to protect yourself. In addition, if most people follow these recommendations, we have chance to diminish the values of parameters α_i and reduce the negative effects of the pandemic.

The parameters ρ_i characterize the patient removal rates, since Eq. (9.3) demonstrates the increase rate of R. The inverse values $1/\rho_i$ are the estimations for time of spreading infection τ_i during ith epidemic wave. So, we are interested in increasing the values of parameters ρ_i and decreasing $1/\rho_i$. People and public authorities should work on this and organize immediate isolation of suspicious cases.

Since the derivative $\mathrm{d}(S+I+R)/\mathrm{d}t$ is equal to zero (it follows from summarizing Eqs. (9.1)–(9.3)), the sum

$$N_i = S + I + R \tag{9.4}$$

must be constant for every wave and is not the volume of population as mentioned in Chap. 4.

To determine the initial conditions for the set of Eqs. (9.1)–(9.3), let us suppose that at the beginning of every epidemic wave t_i^*:

$$I(t_i^*) = I_i,\ R(t_i^*) = R_i,\ S(t_i^*) = N_i - I_i - R_i \tag{9.5}$$

In particular, when the first wave of the epidemic starts with one infected person, the initial conditions (9.5) can be written as follows:

$$I(t_1^*) = 1,\ R(t_1^*) = 0,\ S(t_1^*) = N_1 - 1. \tag{9.6}$$

Initial conditions (9.6) were used in Chap. 6 to simulate the first epidemic waves in different countries.

It follows from (9.1) and (9.2) that

$$\frac{\mathrm{d}I}{\mathrm{d}S} = \frac{v_i}{S} - 1,\ v_i = \frac{\rho_i}{\alpha_i} \tag{9.7}$$

Integration of (9.7) with the initial conditions (9.5) yields:

$$I = v_i \ln S - S + N_i - R_i - v_i \ln(N_i - I_i - R_i) \tag{9.8}$$

It follows from (9.8) that function I has a maximum at $S = v_i$ and tends to zero at infinity. The corresponding number of susceptible persons at infinity $S_{i\infty} > 0$ can be calculated from a nonlinear equation

$$S_{i\infty} = (N_i - I_i - R_i)e^{\frac{S_{i\infty} - N_i - R_i}{v_i}} \tag{9.9}$$

Formula (9.9) follows from (9.8) at $I = 0$.

As in Chap. 4, we have solved (9.1)–(9.3) by introducing the function

$$V(t) = I(t) + R(t), \tag{9.10}$$

corresponding to the number of victims or cumulative confirmed number of cases. For many epidemics (including COVID-19 pandemic), we cannot observe dependencies $S(t), I(t)$ and $R(t)$ but observations of the accumulated number of cases V_j corresponding to the moments of time t_j provide information for direct assessments of the dependence $V(t)$.

It follows from (9.2) and (9.3) that:

$$\frac{dV}{dt} = \alpha_i SI \tag{9.11}$$

Equations (9.4), (9.8) and (9.11) yield:

$$\frac{dV}{dt} = \alpha_i (N_i - V)[v_i \ln(N_i - V) + V - R_i - v_i \ln(N_i - R_i - I_i)] \tag{9.12}$$

Integration of (9.12) provides an analytical solution for the set of Eqs. (9.1)–(9.3):

$$F_i^*(V, N_i, I_i, R_i, v_i) = \alpha_i (t - t_i^*), \tag{9.13}$$

$$F_i^* = \int\limits_{R_i + I_i}^{V} \frac{dU}{(N_i - U)[v_i \ln(N_i - U) + U - R_i - v_i \ln(N_i - R_i - I_i)]}. \tag{9.14}$$

Thus, for every set of parameters $N_i, I_i, R_i, v_i, \alpha_i$ and a fixed value of V, integral (9.14) can be calculated and a corresponding moment of time can be determined from (9.13). Then, functions $I(t)$ and $R(t)$ can be easily calculated with the use of formulas (9.8) and:

$$S = N_i - V, \; R = V - I \tag{9.15}$$

The final numbers of victims (final accumulated number of cases corresponding to the ith epidemic wave) can be calculated from:

$$V_{i\infty} = N_i - S_{i\infty}. \tag{9.16}$$

To estimate the final day of the ith epidemic wave, we can use the condition:

$$I(t_{if}) = 1. \tag{9.17}$$

which means that at $t > t_{if}$ less than one person still spreads the infection.

We have already seen in the previous chapter that not every wave ends with a small number of infected people, so to assess the accuracy of modeling a particular epidemic wave, we will consider its final time point t_{ie} as the moment when the characteristics of the epidemic and SIR parameters changed again. Comparing the calculated value $V(t_{ie})$ with the real accumulated number of cases V_{ie} corresponding t_{ie}(or smoothed values, e.g., (8.1)), we can estimate the accuracy with the use of formula:

$$\varepsilon_i = \frac{|V_{ie} - V(t_{ie})|}{V_{ie}} \tag{9.18}$$

Equation (9.18) is a generalization of (4.20). We have already used it in Chap. 6 to estimate the accuracy of SIR simulations for the first epidemic waves in different countries.

Formulas (4.21) and (4.22) for the probabilities of meeting infected people are valid for every epidemic wave. According to (4.21), the maximum probabilities of meeting an infected person $p(t)$ correspond to the maximum I_{max} of the function $I(t)$. This maximum can be reached at time moments t_i^*, t_{ie} or during the ith epidemic wave $t_i^* < t_{max} < t_{ie}$. In the latter case, the maximum corresponds to $S = v_i$ (see (9.7)). Then, Eq. (9.8) yields:

$$I_{max} \equiv I(t_{max}) = v_i \ln \frac{v_i}{N_i - R_i - I_i} + N_i - v_i - R_i \tag{9.19}$$

The corresponding time moment ($t_i^* < t_{max} < t_{ie}$) can be determined by integration of (9.1)

$$t_{max} = t_i^* - \frac{1}{\alpha} \int\limits_{N_i - I_i - R_i}^{v_i} \frac{\mathrm{d}U}{U[v_i \ln U - U + N_i - R_i - v_i \ln(N_i - I_i - R_i)]} \tag{9.20}$$

In formula (9.20), the initial condition (9.5) and Eq. (9.8) were used.

SIR model also allows us to determine the point in time t_s when the daily increase in new cases will begin to decline. The average daily number of new cases can be estimated as $\mathrm{d}V/\mathrm{d}t$ and calculated with the use of (9.11) and (9.12). Then, the maximum in daily number of new cases corresponds to zero value of the second derivative $\mathrm{d}^2V/\mathrm{d}t^2$. Differentiation of (9.11) yields:

$$\frac{\mathrm{d}^2V}{\mathrm{d}t^2} = \alpha_i\left(S\frac{\mathrm{d}I}{\mathrm{d}t} + I\frac{\mathrm{d}S}{\mathrm{d}t}\right) \tag{9.21}$$

With the use of (9.1) and (9.2), Eq. (9.21) can be rewritten as follows:

$$\frac{\mathrm{d}^2V}{\mathrm{d}t^2} = \alpha_i^2 SI(S - I - v_i) \tag{9.22}$$

According to (9.22), the maximum of the new daily cases corresponds to the value:

$$S_s = I_s + v_i \tag{9.23}$$

(provided it occurs between moments of time t_i^* and t_{ie}). Corresponding value of I_s may be calculated from (9.8):

$$I_s = v_i \ln S_s - S_s + N_i - R_i - v_i \ln(N_i - I_i - R_i) \tag{9.24}$$

Because function S decreases monotonically (see (9.1)), and the maximum of I corresponds to $S = v_i$, Eq. (9.23) shows that the daily number of new cases starts to decline before the maximum number of the infected (and spreading) persons is achieved. This conclusion is very important, because watching a new epidemic, we record the daily number of new cases or the accumulated number of cases. If the daily amount begins to decline, it does not mean that the number of carriers of the infection is also declining and quarantine can be relaxed.

Let us estimate the time difference $t_{\max} - t_s$ (provided both time moments occur between t_i^* and t_{ie}). Equations (9.23) and (9.24) yield the following nonlinear equation for S_s:

$$S_s = \frac{1}{2}\left(v_i + v_i \ln\frac{S_s}{N_i - R_i - I_i} + N_i - R_i\right) \tag{9.25}$$

After solving (9.25), the point in time t_s can be found by integration of (9.1)

$$t_s = t_i^* - \frac{1}{\alpha}\int\limits_{N_i - I_i - R_i}^{S_s} \frac{\mathrm{d}U}{U[v_i \ln U - U + N_i - R_i - v_i \ln(N_i - I_i - R_i)]} \tag{9.26}$$

In Eq. (9.26), the initial condition (9.5) and Eq. (9.8) were used. Formulas (9.20) and (9.26) allow estimating the time difference $t_{\max} - t_s$, but values of the SIR parameters have to be calculated first.

The effective reproduction number $R_{ti}(t)$ shows the average number of people infected by one person during the ith epidemic wave. Similar to (4.32), it can be estimated as:

$$R_{ti}(t) = \frac{I(t+0.5\tau_i) - I(t-0.5\tau_i)}{I} = \frac{1}{I}\int\limits_{t-0.5\tau_i}^{t+0.5\tau_i} \frac{dI(t^*)}{dt^*}dt^* \approx \frac{\tau_i}{I}\frac{dI(t)}{dt} \tag{9.27}$$

where τ_i is the average time of spreading the infection and can be estimated with the use of

$$\tau_i \approx \frac{1}{\rho_i} \tag{9.28}$$

With the use of (9.2), (9.7) and (9.28), Eq. (9.27) can be rewritten as follows:

$$R_{ti}(t) \approx \frac{S}{v_i} - 1 \tag{9.29}$$

The basic reproduction number R_{0i} is used for the initial periods of an epidemic outbreak or for initial periods of consequent epidemic waves when the exponential growth of the number of cases occurs. This characteristic can be estimated with the use of equation similar to (4.33):

$$R_{0i} = \exp\left(\frac{\tau_i \ln 2}{\tau_d}\right) \tag{9.30}$$

where τ_d is the doubling time (the period during which the number of cases duplicates).

Chapter 10
Procedures of Parameter Identification for the Waves of Epidemics

In this chapter, we will discuss the methods of SIR parameter identification for second and next waves of epidemics. Some examples will be presented.

In the case of a new epidemic, the values of its parameters are unknown and must be identified with the use of limited data sets. For the first wave of an epidemic starting with one infected person, the number of unknown parameters is only four, since $I_1 = 1$ and $R_1 = 0$. The corresponding statistical approach was proposed in Chaps. 5 and 6 to estimate the values of four unknown parameters for the first waves of the COVID-19 pandemic in different countries and regions. Some results were published in [35, 38, 66, 67, 69–77].

For the next epidemic waves ($i > 1$), the moments of time t_i^* corresponding to their beginning are known. Therefore, the exact solution (9.13)–(9.14) depends only on five parameters—$N_i, I_i, R_i, v_i, \alpha_i$. Then, the registered number of victims V_j corresponding to the moments of time t_j can be used in Eq. (9.14) in order to calculate $F_{i,j} = F_i^*(V_j, N_i, v_i, I_i, R_i)$ for every fixed values of N_i, v_i, I_i, R_i and then to check how the registered points fit the straight line (9.13).

Equation (9.13) can be rewritten as follows:

$$y \equiv F_i^*(V, N_i, I_i, R_i, v_i) = \alpha_i t - \alpha_i t_i^*, \tag{10.1}$$

Equation (10.1) coincides with (2.2), if we assume

$$\gamma = \alpha_i, \; \beta = -\alpha_i t_i^* \tag{10.2}$$

As in Chap. 2, we can estimate the values of parameters γ and β, by treating the values $y_j \equiv F_i^*(V_j, N_i, I_i, R_i, v_i)$ and corresponding time moments t_j as random variables, and use the observations of the accumulated number of cases and the linear regression in order to calculate the coefficients $\widehat{\gamma}$ and $\widehat{\beta}$ of the regression line

$$\widehat{y} = \widehat{\gamma} t + \widehat{\beta} \tag{10.3}$$

I. Nesteruk, *COVID-19 Pandemic Dynamics*,
https://doi.org/10.1007/978-981-33-6416-5_10

using the standard formulas (2.6) and (2.7). Values $\widehat{\gamma}$ and $\widehat{\beta}$ can be treated as statistics-based estimations of parameters γ and β from relationships (2.3). Then, (10.2) yields the estimations for parameters α_i.

The reliability of the method can be checked by calculating the correlation coefficients r_i for every epidemic wave with the use of (2.8) and checking how close is its value to unity. We can use also the F-test for the null hypothesis that says that the proposed linear relationship (10.1) fits the data set. The experimental values of the Fisher function can be calculated for every epidemic wave with the use of the formula:

$$F_i = \frac{r_i^2(n_i - m)}{(1 - r_i^2)(m - 1)}, \tag{10.4}$$

where n_i is the number of observations for the ith epidemic wave and $m = 2$ is the number of parameters in the regression equation. The corresponding experimental value F_i has to be compared with the critical value $F_C(k_1, k_2)$ of the Fisher function at a desired significance or confidence level ($k_1 = m - 1$, $k_2 = n_i - m$). When the values n_i and m are fixed, the maximum of the Fisher function coincides with the maximum of the correlation coefficient. Therefore, to find the optimal values of parameters N_i, v_i, I_i, R_i, we have to find the maximum of the correlation coefficient for the linear dependence (10.1). To compare the reliability of different predictions (with different values of n_i), it is useful to apply the ratio $F_i/F_C(1, n_i - 2)$ at a fixed significance level. We will use the level 0.001; corresponding values of $F_C(1, n_i - 2)$ can be taken from [34]. The most reliable prediction yields the highest $F_i/F_C(1, n_i - 2)$ ratio.

The exact solution (9.13)–(9.14) allows avoiding numerical solutions of differential Eqs. (9.1)–(9.3) and significantly reduces the time spent on calculations. For large values of V_j (e.g., in the case of global dynamics), calculations of integral (9.14) need more computer time, but even in these cases the calculation of one set of optimal parameters could be performed on a regular laptop for one day. We have modified our MATLAB codes (see Chap. 5) in order to perform calculations of the next pandemic waves.

In the case of sequential calculation of epidemic waves i = 1, 2, 3 …, it is possible to avoid determining the four optimal unknown parameters N_i, v_i, I_i, R_i, thereby reducing the amount of calculations and difficulties in isolation of a maximum of correlation coefficient. For parameters I_i, R_i, it is possible to use the numbers of I and R calculated for the previous wave of epidemic at the moment of time when the following wave began. Then, we need to calculate values $F_i^*(V_j, N_i, v_i)$, linear regression coefficients (10.3), correlation coefficient r_i, $F_i/F_C(1, n - 2)$ and to isolate the values of parameters N_i and v_i corresponding to the maximum of the correlation cofficient. The MATLAB code mentioned in Chap. 5 required only minor modifications. Knowing the optimal values of five parameters $N_i, I_i, R_i, v_i, \alpha_i$, the SIR curves and other characteristics of the corresponding epidemic wave can be calculated with the use of formulas presented in previous chapter. A separate original MATLAB code was used for these calculations. This approach has been successfully used in [89, 94]. In particular, six waves of the COVID-19 epidemic in Ukraine and four pandemic waves in the world were calculated. Corresponding results will be presented in the next two sections.

Segmentation of epidemic waves and their sequential SIR simulations need a lot of efforts. To avoid this, a new method of obtaining the optimal values of SIR parameters can be proposed. First of all, we can use the relationship

$$V_i = I_i + R_i \tag{10.5}$$

which follows from (9.10). To estimate the value V_i, we can use the smoothed accumulated number of cases (e.g., formula (8.1)). Then,

$$V_i \approx \frac{1}{7}\sum_{j=i-3}^{j=i+3} V_j \tag{10.6}$$

where i corresponds to the moment of time t_i^*. To obtain one more relationship, let us use (9.4), (9.10) and (9.11)

$$I_i = \frac{1}{\alpha_i(N_i - V_i)}\frac{\mathrm{d}V}{\mathrm{d}t}\Big|_{t=t_i^*} \tag{10.7}$$

To estimate the average number of new cases $\mathrm{d}V/\mathrm{d}t$ at the moment of time t_i^*, we can use (8.2). Thus, we have only two independent parameters N_i and v_i, but we do not have an explicit formula to calculate parameter α_i. Its value can be obtained with the use of iterations. The first approximation can be taken from formula (7.17):

$$\alpha_i^{(1)} = 6.35645e - 07N_i^{-0.9352} \tag{10.8}$$

Next approximations can be calculated with the use of (10.2) and (2.7) as follows:

$$\alpha_i^{(k)} = -\frac{\widehat{\beta}}{t_i^*}, \quad k = 2, 3, 4, \ldots \tag{10.9}$$

In the case of convergence, the relative error

$$\varepsilon_\alpha^{(k)} = \frac{\left|\alpha_i^{(k)} - \alpha_i^{(k-1)}\right|}{\alpha_i^{(k)}}, \quad k = 2, 3, 4, \ldots \tag{10.10}$$

tends to zero with increasing number of iteration k.

The suitability of the method was tested with the use of Ukrainian data set, presented in Table 8.2 (number of observations n = 14, T_c: October 1–14).

The convergence was rather good (after 10 iterations, the values $\varepsilon_{\alpha}^{(k)}$ were less than 10^{-4} for some ranges of independent parameters N_i and v_i). The calculations were performed with the use of original MATLAB code. Knowing the optimal values of the parameters N_i, v_i, I_i, R_i, anyone familiar with the differential equations can obtain SIR curves and predictions with the use of formulas available in the previous chapter.

An example of using presented algorithm is shown in Table 10.1 and in Fig. 10.1. The calculations of the first epidemic wave in Ukraine can be found in Sect. 6.11. Waves 2–6 were calculated in [89, 94] and will be discussed in the next chapter. That is why we use the name of seventh wave for this SIR simulation.

Table 10.1 demonstrates very high values of the correlation coefficient (close to unit) and high values of the $F_i/F_C(1, n_i - 2)$ ratio. The final size of this epidemic wave in Ukraine $V_{7\infty}$(around 400 thousands) is much higher than in previous estimations; nevertheless, it looks still too optimistic, since the real values of V_j registered after T_c (see "stars" in Fig. 10.1) are higher than the theoretical estimation (solid line). According to this prediction, the epidemic cannot stop before late March 2021. The average spreading time $\tau_7 = 1/\rho_7$ was estimated as 5.4 days. This figure is slightly higher than it was during the first wave (4 days, see Table 6.23, prediction 8). This fact can be explained by lockdown applied during the first epidemic wave in Ukraine.

Table 10.1 Calculated optimal values of SIR parameters for the seventh wave of the COVID-19 epidemic in Ukraine

Characteristics	Ukraine, seventh wave, $i = 7$, $n = 14$
Period taken for calculations T_c	October 1–14
I_i	17,605.4705407746
R_i	200,051.672316368
N_i	566,602.33216
v_i	271,873.891638209
α_i	6.81143662034146e-07
$R_{ti}(t_i^*)$, Eq. (9.29)	0.28
t_i^*	255
ρ_i	0.185185178161924
$1/\rho_i$	5.40000020479830
r_i	0.999637260478932
F_i, Eq. (10.4)	16,531.7953887678
$F_i/F_C(1, n_i - 2)$	888.806203697193
$S_{i\infty}$, Eq. (9.9)	168,285
$V_{i\infty}$, Eq. (9.16)	398,317
t_{if}, Eq. (9.17)	432
Final day of the seventh wave	March 27, 2021

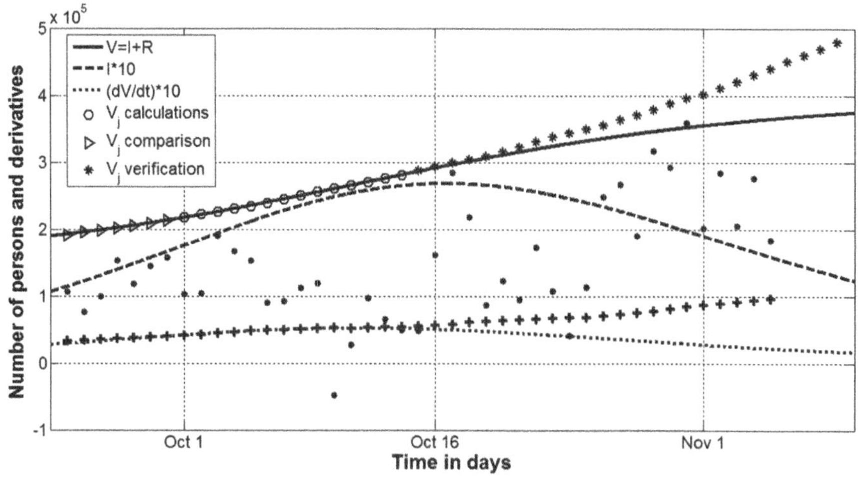

Fig. 10.1 Seventh and next epidemic waves in Ukraine: SIR curves (lines) and accumulated number of cases (markers) versus time. Numbers of victims $V = I + R$ (solid), infected and spreading I multiplied by 10 (dashed), daily new cases dV/dt (Eq. (9.10)) multiplied by 10 (dotted); "circles" represent the V_j values taken for calculations of the seventh wave; "triangles" show the accumulated numbers of cases before T_c; "stars"—after T_c. "Crosses" show the first derivative of the smoothed accumulated number of cases (Eq. (8.2)) multiplied by 10, "dots"—the second derivative (Eq. (8.3)) multiplied by 1000

Figure 10.1 demonstrates that the calculated $V(t)$ curve (solid line) fits the points taken for calculations ("circles") and some points before T_c ("triangles"). But observations of the epidemic dynamics after T_c ("stars") demonstrate that real V_j values deviate rather quick from the theoretical solid curve. It could be the result of the changes in epidemic parameters. To check this hypothesis, the methods of epidemic wave detection can be used (see Chap. 8). Formulas (8.2), (8.3) and data sets from Tables 8.2 and 10.2 were used for calculations. The results are also shown in Fig. 10.1 by "crosses" (first derivative (8.2) multiplied by 10) and "dots" (second derivative (8.3) multiplied by 1000).

By October 16, we see a very good correspondence between the registered daily number of new cases (Eq. (8.2), "crosses") and the theoretical estimate (Eq. (9.11), dotted line). The differences began after October 16 at the same time as the jump of the second derivative (see "dots" in Fig. 10.1). Probably, this is due to the preparations for the local elections and a presidential poll, which were held throughout Ukraine on October 25, 2020, and involved hundreds of thousands of people to campaign and work in election commissions (their number was about 30 thousand). This obviously increased the number of contacts and the likelihood of additional infections.

Looking at the dashed curve in Fig. 10.1, we can conclude that in late September, the number of spreaders of infection exceeded 15 thousands. Formula (4.21) yields the corresponding probability $p \approx 3.4 \cdot 10^{-4}$ to meet such a person in

Table 10.2 Official cumulative numbers of confirmed COVID-19 cases in Ukraine [95, 96]

Day in October and November 2020	Time in days t_j	Accumulated number of cases in Ukraine V_j
19	373	309,107
20	374	315,826
21	375	322,879
22	376	330,396
23	377	337,410
24	378	343,498
25	379	348,924
26	380	355,601
27	381	363,075
28	382	370,417
29	383	378,729
30	384	387,481
31	385	395,440
1	386	402,194
2	387	411,093
3	388	420,617
4	389	430,467
5	390	440,188
6	391	450,934
7	392	460,331
8	393	469,018
9	394	479,197

Ukraine. If someone had 100 additional contacts, the probability of meeting at least one infected person is approximately 0.03 (see Eq. (4.22)). There is also a visible jump in second derivative on October 25 (see “dots” in Fig. 10.1).

Further consequences of election holding will be noticeable in November. But already on October 31, we see the highest value of the second derivative (“dots” in Fig. 10.1). Thus, we can conclude that elections and the presidential poll caused the new epidemic waves in Ukraine and a significant increase in the number of cases. To have reliable predictions for Ukraine, we have to continue calculations taking into account new epidemic waves. In particular, we can assume that the eighth epidemic wave had started on October 31. Some approximate estimation for the final size of this wave and its duration will be presented in Chap. 13.

We must understand also that different parameter identification procedures can yield different optimal values of the SIR model parameters (see, e.g., [56]). The fastest verification method is to compare calculated SIR curves for $V = I + R$ with V_j values, obtained after the calculations have been completed. It is possible also to use dV/dt curve (Eq. (9.11), dotted line in Fig. 10.1). For example, corresponding dotted line in Fig. 10.1 demonstrates that according to the prediction of seventh epidemic wave in Ukraine, the average daily number of new cases has to diminish after October 9. The smoothed real numbers (shown by dots) started to diminish simultaneously, but this tendency finished after October 11. Other examples of SIR simulations for Ukraine will be presented in the next chapter.

Chapter 11
Applications of the General SIR Model for Calculations of the COVID-19 Epidemic Waves in Ukraine

In this chapter, we will use the general SIR model to simulate five COVID-19 epidemic waves in Ukraine. The results of calculations and corresponding predictions will be presented.

The generalized SIR model allows us to simulate new waves of an epidemic caused by changes in quarantine conditions, social behavior, pathogen activity, etc. In previous chapter, two algorithms of the model parameter identification were proposed. Here, we consider the case of sequential calculation of epidemic waves $i = 1, 2, 3 \ldots$, when it is possible to avoid determining the four optimal unknown parameters N_i, v_i, I_i, R_i since the values of I_i and R_i can be taken from dependences for I and R calculated for the previous wave of epidemic at the moment of time when the following wave began. This approach has been successfully used in [89, 94]. In this chapter, five COVID-19 epidemic waves in Ukraine will be calculated with the use of general SIR model developed in Chaps. 9 and 10.

We will use the data sets for accumulated number of cases V_j in Ukraine presented in Tables 6.22, 8.1 and 8.2 and the results of SIR simulation of the first wave (prediction 8) presented in Chap. 6. The moments of time when the epidemic dynamics changed (and new waves began) were identified in Chap. 8. The number of observations was the same for all cases ($n_i = 14$; $i = 2, 3, 4, 5, 6$). The corresponding time periods T_{ci} used for calculations, optimal values of parameters and other characteristics of the epidemic waves are shown in Tables 11.1 and 11.2.

All the predictions shown in Tables 10.1, 11.1 and 11.2 are too optimistic since the accumulated number of cases exceeded 400 thousand already on November 1, 2020. It means that even the seventh epidemic wave (calculated in the previous chapter) is not the last one, unfortunately. Permanent changes in epidemic dynamics will cause high values of the relative accuracy ε_i (it was impossible to apply formula (9.18) in November 2020, and corresponding values are not shown in Tables 10.1, 11.1 and 11.2), despite rather high values of $F_i/F_C(1, n-2)$ and the correlation coefficients r_i. For the first epidemic wave in Ukraine, we have calculated $\varepsilon_1 = 89.3\%$ (see Table 6.23, prediction 8).

I. Nesteruk, *COVID-19 Pandemic Dynamics*,
https://doi.org/10.1007/978-981-33-6416-5_11

Table 11.1 Characteristics of second, third and fourth epidemic waves in Ukraine. Optimal values of parameters, final sizes and days (last three rows)

Characteristics	Ukraine, second wave, $i = 2$	Ukraine, third wave, $i = 3$	Ukraine, fourth wave, $i = 4$
Period taken for calculations, T_{ci}	May 17–30	June 9–22	July 3–16
I_i	2,196	2,490	3,736
R_i	16,420	25,891	43,941
N_i	239,587.2	149,444.16	182,294.495539200
v_i	209,382.9208704	104,004.527376384	119,723.766724183
α_i	7.728220925e-07	2.117564829222e-06	1.35741149806184e-06
$R_{ti}(t_i^*)$, Eq. (11.1)	0.13	0.16	0.12
t_i^*	118	141	165
ρ_i	0.16181574704683	0.220236329252125	0.162514417542680
$1/\rho_i$	6.17986826529682	4.54057694929708	6.15330021250195
r_i	0.99889478036988	0.999714332158213	0.999456710370030
F_i, Eq. (10.4)	5,419.78688520125	20,994.4146122108	11,034.8339181942
$F_i/F_C(1, n_i - 2)$	291.386391677487	1,128.73196839843	593.270640763128
$S_{i\infty}$, Eq. (9.9)	178,641	78,509	89,552
$V_{i\infty}$, Eq. (9.16)	60,946	70,935	92,742
t_{if}, Eq. (9.17)	531.9	338.5	421.7
Final day of ith epidemic wave	July 5, 2021	December 24, 2020	March 17, 2021

Figure 11.1 illustrates first four waves of the COVID-19 epidemic in Ukraine shown by different colors (red, black, blue and magenta for waves 1 (prediction 8), 2, 3 and 4, respectively). SIR curves represent the numbers of infected and spreading $I \times 10$ (dashed lines) and the number of victims $V = I + R$ (solid lines). “Circles” in corresponding colors represent the values V_j taken for calculations, and “stars” show the accumulated number of cases used only for the verification of calculations. Every new wave increased the estimation of the final size $V_{i\infty}$ of the epidemic in Ukraine (see also Tables 11.1 and 11.2).

Every new wave also increases the number of infected and spreading the infection persons I (represented by dashed lines). The accuracy of simulations is rather good, especially for second and further pandemic waves, since blue “stars” (showing the accumulated values of confirmed cases used only for verification of the calculations) are located very close to the corresponding solid lines showing the calculated $V = I + R$ values. There are some discrepancies for the early stages of the first wave, when the number of registered cases is lower than the real one due to the problems with the identification of the infected persons. It can be seen also in the beginning of the fifth wave in Ukraine. After July 23, the real number of cases is

Table 11.2 Characteristics of fifth and sixth epidemic waves in Ukraine. Optimal values of parameters, final sizes and days (last three rows)

Characteristics	Ukraine, fifth wave, $i = 5$	Ukraine, sixth wave, $i = 6$
Period taken for calculations, T_{ci}	July 20–August 2	August 9–22
I_i	4,570	7,525
R_i	55,596	73,424
N_i	208,624.896	336,210.688000000
v_i	117,860.44860629	166,678.304458998
α_i	1.241234878e-06	6.47063569e-07
$R_{ti}(t_i^*)$, Eq. (11.1)	0.26	0.53
t_i^*	182	201
ρ_i	0.1462924996005	0.107851458558717
$1/\rho_i$	6.83562043666339	9.27201183334553
r_i	0.9997539353297	0.999591012582638
F_i, Eq. (10.4)	24,374.834072843	14,661.3785892930
$F_i/F_C(1, n_i - 2)$	1,310.47495015287	788.246160714677
$S_{i\infty}$, Eq. (9.9)	79,668	91,157
$V_{i\infty}$, Eq. (9.16)	128,957	245,054
t_{if}, Eq. (9.17)	381	443.5
Final day of i-th epidemic wave	February 5, 2021	April 8, 2021

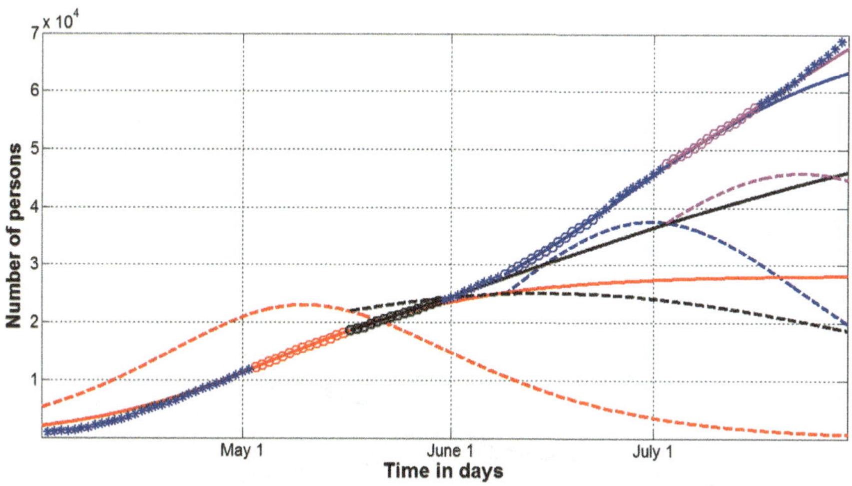

Fig. 11.1 Four COVID-19 epidemic waves in Ukraine. SIR curves (lines) and accumulated number of cases (markers) versus time. Red, black, blue and magenta colors correspond to waves 1 (prediction 8), 2, 3 and 4, respectively. Numbers of infected and spreading $I \times 10$ (dashed lines) and victims $V = I + R$ (solid lines). "Circles" represent the values V_j taken for calculations, and "stars" show the accuracy of calculations

higher than the calculated one for the fourth wave (compare blue “stars” and solid magenta line in Fig. 11.1).

Figure 11.2 illustrates first six waves of the COVID-19 epidemic in Ukraine shown by different colors (red, black, blue magenta, green and brown) for waves 1 (prediction 8), 2, 3, 4, 5 and 6, respectively. SIR curves represent the numbers of infected and spreading $I \times 10$ (dashed lines) and the number of victims $V = I + R$ (solid lines). “Circles” in corresponding colors represent the values V_j taken for calculations, and “stars” show the accumulated number of cases used only for the verification of calculations. It can be seen the estimations of the final size $V_{i\infty}$ for the new fifth, sixth and seventh (simulated in previous chapter) continue to increase (see also Tables 10.1, 11.1 and 11.2).

Every new wave also increases the number of infected and spreading the infection persons I (represented in Figs. 11.1 and 11.2 by dashed lines). A particularly sharp increase in the number of people spreading the infection is characteristic of the sixth wave (after the beginning of August), which is explained by the holiday season, during which the mobility of the population increased and there was a mass disregard for social distance.

The accuracy of simulations looks rather good, especially for the six wave, since more than one month the registered number of cases (V_j blue “stars”) did not deviate from the theoretical $V = I + R$ curve (solid brown line). The discrepancies started to appear after mid-September and are connected with the beginning of the academic year on September 1. In the Ukrainian regions with the largest daily increase in new cases, it was advisable to use only a distancing form of education.

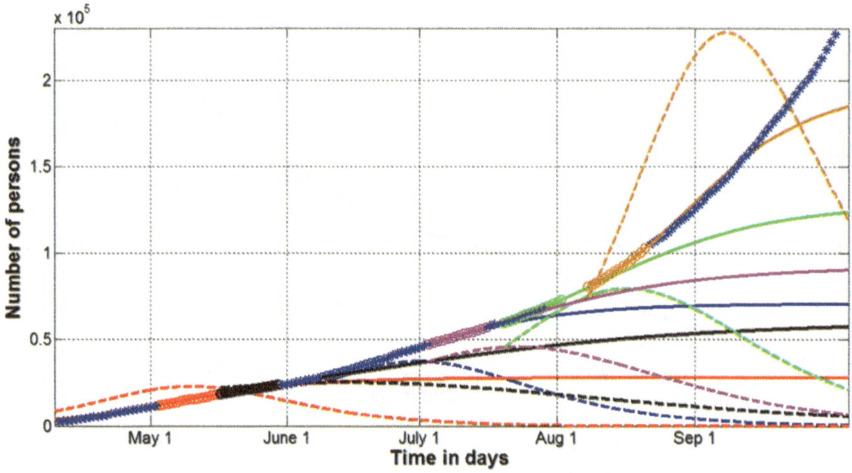

Fig. 11.2 Six COVID-19 epidemic waves in Ukraine. SIR curves (lines) and accumulated number of cases (markers) versus time. Red, black, blue, magenta, green and brown colors correspond to waves 1 (prediction 8), 2, 3, 4, and 6, respectively. Numbers of infected and spreading $I \times 10$ (dashed lines) and victims $V = I + R$ (solid lines). “Circles” represent the values V_j taken for calculations, and “stars” show the accuracy of calculations

Figure 11.3 illustrates the changes in SIR parameters and epidemic characteristics versus the number of epidemic wave. The increase in the number of waves i corresponds to later periods of time used for calculations T_{ci}. We can see monotonous increase in the predicted final sizes $V_{i\infty}$ (small "triangles") and almost monotonous (second wave is an exception) in N_i values ("circles"). The second wave is also an exception in monotonous decrease in α_i (small "stars"). Probably, this peculiarity can be explained with rather chaotic epidemic dynamics during T_{c2} (May 17–30) visible in Fig. 8.2 (see jumps of second derivative shown by "stars"). In this period main restrictions (in particular, connected with transportation) have been lifted or relaxed (the national lockdown was canceled on May 10, 2020). For the second wave, we can see also higher values of parameter τ in comparison with the previous epidemic wave ($\tau_2 > \tau_1$) and the next one ($\tau_2 > \tau_3$). It means that average period of spreading the infection was longer during the second wave. But the highest values τ_i correspond to the holiday time (fifth and sixth epidemic waves). Probably, many infected people traveled and spread the infection longer in comparison with the periods before and after July–August 2020.

The predicted final sizes $V_{i\infty}$ can be compared with the approximation (7.18) (large "triangles" in Fig. 11.3). You can see a pretty good correspondence, despite the fact that the formula (7.18) was obtained by systematizing the results of calculations only for the first pandemic waves. A comparison of the values α_i (small "stars") with the values that can be obtained by applying formula (7.17) (large "stars") shows a slightly worse fit. But as the experience of calculating the seventh wave has shown, approximation (7.17) can be a good first approximation of the value α_i in the iterative procedure described in Chap. 10.

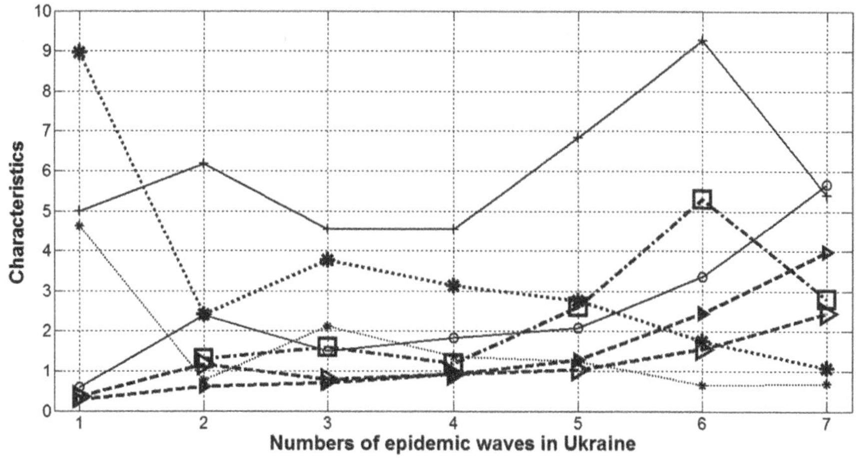

Fig. 11.3 SIR parameters for different COVID-19 epidemic waves in Ukraine: "circles" represent $N_i/100000$; "triangles"—$V_{i\infty}/100000$ (large markers correspond to approximation (7.18)); "stars"—$\alpha_i \cdot 10^6$ (large markers correspond to approximation (7.17)); "crosses"—$\tau_i = 1/\rho_i$; "squares"—$R_{ti}(t_i^*) \cdot 10$, formula (11.1)

We can also easily calculate the effective reproduction numbers at the moments of time t_i^* corresponding to the beginning of the ith wave with the use of (9.4) and (9.29):

$$R_{ti}(t_i^*) \approx \frac{(N_i - I_i - R_i)}{v_i} - 1 \quad (11.1)$$

The results of calculations are shown in Tables 10.1, 11.1 and 11.2 and in Fig. 11.3 (by "squares"). It can be seen that the highest values of the reproduction number correspond to the vacation season.

Chapter 12
Global Waves of the COVID-19 Pandemic

In this chapter, we will use the general SIR model to simulate the global COVID-19 pandemic waves. The results of calculations and corresponding predictions will be presented.

The generalized SIR model (see Chaps. 9 and 10) allows us to simulate new waves of an epidemic caused by changes in quarantine conditions, social behavior, pathogen activity, etc. Here, we consider the case of sequential calculation of epidemic waves $i = 1, 2, 3 \ldots$, when it is possible to avoid determining the four optimal unknown parameters N_i, v_i, I_i, R_i since the values of I_i and R_i can be taken from dependences for I and R calculated for the previous epidemic wave at the moment of time when the following wave began. This approach has been successfully used in [89, 94]. In this chapter, four COVID-19 pandemic waves in the world will be simulated.

We will use the data sets for global accumulated number of cases V_j presented in Table 6.22 and the results of SIR simulation of the first wave presented in Chap. 6 (Table 6.25, prediction 2). The moments of time when the epidemic dynamics changed (and new waves began) were identified in Chap. 8. The number of observations was the same for all the cases ($n_i = 14$; $i = 2, 3, 4, 5$). The corresponding time periods T_{ci} used for calculations, optimal values of parameters and other characteristics of the pandemic waves are shown in Tables 12.1 and 12.2.

All the predictions shown in Tables 6.25, 12.1 and 12.2 are too optimistic since the global accumulated number of cases exceeded 47 millions already on November 2, 2020. It means that the fifth epidemic wave is not the last one, unfortunately. Permanent changes in the pandemic dynamics will cause high values of the relative accuracy ε_i (it was impossible to apply formula (9.18) in November 2020, and corresponding values are not shown in Tables 12.1 and 12.2), despite rather high values of $F_i/F_C(1, n-2)$ and the correlation coefficients r_i. The predictions of the final day of the COVID-19 pandemic waves are very different and not optimistic. For example, the prediction for the second wave (see Table 12.1) yielded only February 2022 as a possible end of the pandemic. It must be noted the final sizes $V_{i\infty}$ predicted for different global waves do not increase monotonously

I. Nesteruk, *COVID-19 Pandemic Dynamics*,
https://doi.org/10.1007/978-981-33-6416-5_12

Table 12.1 Characteristics of the second and third pandemic waves in the world. Optimal values of parameters, final sizes and days (last three rows)

Characteristics	Global second wave, $i = 2$	Global third wave, $i = 3$
Period taken for calculations, T_{ci}	May 12–25	May 28–June 10
I_i	295,912	377,816
R_i	3,874,512	5,323,521
N_i	113,600,000	95,388,160
ν_i	102,947,264.512000	88,021,224.0422502
α_i	2.66429099926749e-09	3.5269848715e-09
$R_{ti}(t_i^*)$, Eq. (11.1)	0.06	0.02
t_i^*	113	129
ρ_i	0.274281470238531	0.310449525572027
$1/\rho_i$	3.64588974650873	3.22113553936803
r_i	0.999760778228936	0.999820043845012
F_i, Eq. (10.4)	25,072.3295928952	33,332.4550282053
$F_i/F_C(1, n_i - 2)$	1,347.97470929544	1,792.06747463469
$S_{i\infty}$, Eq. (9.9)	93,209,515	79,959,644
$V_{i\infty}$, Eq. (9.16)	20,390,485	15,428,516
t_{if}, Eq. (9.17)	749.5	652.3
Final day of pandemic wave	February 18, 2022	November 2, 2021

Table 12.2 Characteristics of the fourth and fifth pandemic waves in the world. Optimal values of parameters, final sizes and days (last three rows)

Characteristics	Global fourth wave, $i = 4$	Global fifth wave, $i = 5$
Period taken for calculations, T_{ci}	June 13–26	July 5–18
I_i	392,933	562,468
R_i	7,297,775	10,765,332
N_i	94,880,000	47,113,472
ν_i	81,424,517.3514240	32,637,843.7070307
α_i	3.903409800432e-09	9.51859566656283e-09
$R_{ti}(t_i^*)$, Eq. (11.1)	0.07	0.10
t_i^*	145	167
ρ_i	0.317833259025022	0.310666437675697
$1/\rho_i$	3.14630382945943	3.21888649279808
r_i	0.999654631292158	0.999700105970476
F_i, Eq. (10.4)	17,363.7383017448	19,998.0676446882
$F_i/F_C(1, n_i - 2)$	933.534317298105	1,075.16492713378
$S_{i\infty}$, Eq. (9.9)	72,028,177	26,315,036
$V_{i\infty}$, Eq. (9.16)	22,851,823	20,798,436
t_{if}, Eq. (9.17)	600	427.2
Final day of pandemic wave	September 11, 2021	March 22, 2021

(in comparison with the case of Ukraine). Probably due to the fact that global dynamics are influenced by many more different factors (as we saw in Chap. 7), the situation in different countries and regions is very different.

Figure 12.1 illustrates five COVID-19 pandemic waves in the world shown by different colors (red, black, blue, magenta and brown for waves 1 (prediction 2), 2, 3, 4, and 5, respectively). SIR curves represent the numbers of infected and spreading $I \times 10$ (dashed lines) and the number of victims $V = I + R$ (solid lines). "Circles" in corresponding colors represent the values V_j taken for calculations, and "stars" show the accumulated number of cases used only for the verification of calculations.

The numbers of infected and spreading the infection persons I are represented by dashed lines. The accuracy of simulations is rather good, especially for second, third and forth waves, since blue "stars" (showing the accumulated values of confirmed cases used only for verifications of the calculations) are located very close to the corresponding solid lines showing the calculated $V = I + R$ values. There are some discrepancies for the early stages of the first wave, when the number of registered cases is lower that the real one due to the problems with the identification of the infected persons. Similar can also be seen at the beginning of the sixth pandemic wave. After July 26, the real number of cases is higher than the calculated one for the fifth pandemic wave (compare blue "stars" and solid brown line in Fig. 12.1).

We can use Eq. (4.21) to estimate the probability of meeting an infected person in different countries and regions. With the use of corresponding SIR curves, it is

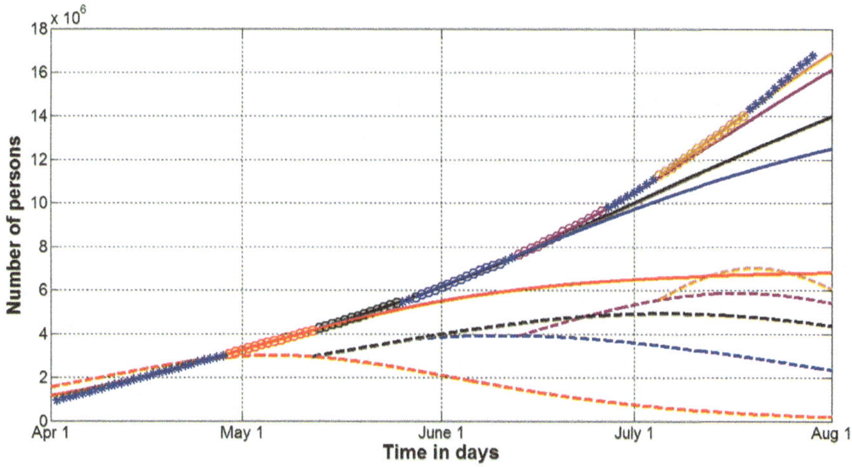

Fig. 12.1 Pandemic waves in the world. SIR curves (lines) and accumulated number of cases (markers) versus time. Red, black, blue, magenta and brown colors correspond to waves 1, 2, 3, 4 and 5, respectively. SIR curves represent the numbers of infected and spreading $I \times 10$ (dashed lines) and victims $V = I + R$ (solid lines). "Circles" represent the values V_j taken for calculations, and "stars" show the accuracy of calculations

possible to estimate $p(t)$ for every moment of time. For example, on July 19, 2020, these probabilities can be estimated as 1.04×10^{-4} and 9.28×10^{-5} for Ukraine and the world, respectively. It means that Ukraine was 12% more dangerous in comparison with the average situation in the world.

Formula (4.21) can be useful for deciding which countries' citizens are welcome guests. If $p_A(t) > p_B(t)$ for countries A and B, citizens of A are not welcome guests in country B at the moment of time t. Since such decisions are very important, we will propose some simple method of estimating $p(t)$ without calculations of SIR parameters for epidemic waves in a country. According to (9.11):

$$I = \frac{1}{\alpha_i S}\frac{\mathrm{d}V}{\mathrm{d}t} \tag{12.1}$$

Since function $S(t)$ changes not very sufficient during a wave (compare initial $S_i = N_i - I_i - R_i$ and final $S_{i\infty}$ values of this function in Tables 12.1 and 12.2 and in corresponding Tables in Chaps. 10 and 11), we can use the constant value S_i instead of $S(t)$. To estimate $\mathrm{d}V/\mathrm{d}t$, it is possible to use Eqs. (8.2) or (8.4). Finally, with the use of (12.1), Eq. (4.21) can be written as follows:

$$p(t) \approx k_p \frac{\mathrm{d}V}{\mathrm{d}t}; \tag{12.2}$$

$$k_p = \frac{1}{S_i \alpha_i N_{\mathrm{pop}}} \tag{12.3}$$

The results of calculations for different countries and regions are presented in Table 12.3. It can be seen that coefficient k_p changes not very fast (compare values on different days for Ukraine and the world). It means that values of this coefficient can be used during rather long period of time before new exact estimations of $I(t)$ will be available. According to Eq. (12.2), the probability of meeting an infected person mostly depends on the daily increase in the number of cases. The values p are estimated for July 19, 2020, and presented in Table 12.3.

The p value for the fourth wave in Ukraine is 20% lower than the exact estimation 1.04×10^{-4} calculated before with the use of Eq. (4.21). In the case of the world fifth wave, the difference between the exact estimation (4.21) and the approximate formulas (12.2) and (12.3) is only 8%. It can be seen that the most infected country was USA. The average global figure was approximately 4 times less. The probabilities p in Kyiv, Moldova and Ukraine were around the world average value. Lower figures were calculated for Spain and the UK. As of July 19, 2020, the safest counties were Italy, France, Austria, and Germany.

The effective reproduction number $R_{ti}(t)$ can be estimated using Eq. (11.1) and constant value $S_i = N_i - I_i - R_i$ instead of $S(t)$. The results are shown in the last column of Table 12.3. It can be seen that France, Italy, Germany, and Austria had the lowest values of R_{ti} which are close to the world figure (wave 5). Not much higher values were calculated for USA and Ukraine (wave 4). Spain, the UK,

Table 12.3 Epidemic characteristics for different countries and regions calculated for different days

Country or region, number of prediction or wave	Day used for calculations	Volume of population N_{pop}	$\alpha_i S_i$	k_p, Eq. (12.3)	Probability p, on July 19, 2020 Eqs. (12.2) and (12.3)	Effective reproduction number R_{ti}, Eq. (11.1)
World, wave 1, prediction 2	April 29	7594×10^6	0.3257	4.04×10^{-10}	1.1×10^{-4}	0.184
World, wave 5	July 5	7594×10^6	0.3406	3.87×10^{-10}	1.0×10^{-4}	0.096
Ukraine, wave 1, prediction 8	May 3	43,716,569	0.2729	8.38×10^{-8}	5.6×10^{-5}	0.363
Ukraine, wave 4	July 3	43,716,569	0.1827	1.25×10^{-7}	8.3×10^{-5}	0.124
USA	April 9	331,002,651	0.7715	3.92×10^{-9}	4.0×10^{-4}	0.122
Germany, 2	April 9	83,783,942	0.8268	1.44×10^{-8}	5.6×10^{-6}	0.116
The UK	April 9	67,886,011	0.4384	3.36×10^{-8}	3.9×10^{-5}	0.326
Italy, 6	March 29	60,461,826	0.7608	2.17×10^{-8}	4.4×10^{-6}	0.116
Spain, 2	March 29	46,754,778	0.8120	2.63×10^{-8}	6.0×10^{-5}	0.182
France, 2	April 5	65,283,190	1.3569	1.13×10^{-8}	7.1×10^{-6}	0.108
Austria, 6	March 28	9,009,755	1.4471	7.67×10^{-8}	6.6×10^{-6}	0.116
Moldova, 2	April 5	4,033,963	0.4646	5.34×10^{-7}	8.6×10^{-5}	0.499
Kyiv, 2	April 11	2,988,176	0.3598	9.30×10^{-7}	1.1×10^{-4}	0.603

Moldova, and Kyiv demonstrate the highest figures. The large difference in Ukrainian and Kyiv values can be explained by very different days of estimation (July 3 and April 11, respectively).

The R_{ti} values corresponding to July 19, 2020, were estimated for Ukraine (wave 4, Table 11.1) and the world (wave 5, Table 12.2) with the use of exact dependences $I(t)$ in formula (9.27) in order to compare with the results presented in Table 12.3. The values $R_t = 0.0257$ and $R_t = 0.0059$ for Ukraine and the world, respectively, demonstrate that the effective reproduction number can decrease very rapidly over time.

The effective reproduction numbers of the COVID-19 pandemic were estimated with the use of different mathematical models for different countries [16, 18, 24, 30–32]. In particular, in [32] the R_t values were estimated as of May 10 for EU countries. Corresponding values for Austria, Spain, Germany, France and Italy vary from 0.45 to 0.74 and significantly exceed those shown in Table 12.3. Rapid changes in R_t values for the epidemic in China are reported in [16] (from 2.35 on January 16 to 1.05 on January 31).

Chapter 13
Long-Time Predictions for the Pandemic Dynamics

In this chapter, we will discuss the problems of long-time predictions of the COVID-19 pandemic dynamics with the use of mathematical models. Some approximate formulas will be proposed in order to estimate the final size and duration of the epidemic with the use of limited data sets at the beginning of new waves. Prediction for the world dynamics, USA, Germany, the UK, the Republic of Korea, Italy, Austria, Spain, France, Sweden, the Republic of Moldova, Ukraine, and the city of Kyiv will be presented.

Any mathematical modeling of the epidemic dynamics will be of particular value if we make an accurate long-term forecast of its duration and number of diseases using statistics data sets obtained immediately after the outbreak. That is why many authors were trying to predict the COVID-19 pandemic dynamics in many countries and regions [7, 11, 17, 19–32, 38–51, 57–78, 81, 83, 89, 90, 94, 97–134]. Unfortunately, the correct mathematical simulation of the COVID-19 pandemic is very difficult for at least two reasons.

First, data on the number of cases were clearly incomplete immediately after outbreaks, and there were quite long hidden periods (we discussed this problem in previous chapters; see, for example, Fig. 6.29). The reason is the large number of asymptomatic patients and the lack of skills to detect a new disease. If the symptoms are obvious (as was, for example, in the case of the mysterious children illness in the Ukrainian city of Chernivtsi), the number of reported cases immediately after the epidemic outbreak may correspond to the actual number and allow adequate long-term prognosis [35].

It must be noted that large discrepancy between registered and actual number of cases occurred even for later periods of the COVID-19 pandemic. Total testing in Slovakia (65.5% of population tested on October 31–November 1, 2020) revealed a number of previously undetected cases, equal to about 1% of the population [135]. According to the WHO report at the end of October, the number of detected cases was also approximately 1% of the country's population [1]. Thus, even six months after the detection of the first patient, half of the cases remained undetected. Probably, we have a similar situation not only in Slovakia.

I. Nesteruk, *COVID-19 Pandemic Dynamics*,
https://doi.org/10.1007/978-981-33-6416-5_13

Relatively small number of people who spread COVID-19 infection can cause a fairly long course of the epidemic. In particular, 20 such persons in Ukraine (estimation at the beginning of 2021, see Fig. 13.1) can cause new cases for another two and a half months (see Table 11.1, wave 4). This explains the many new epidemic waves in Europe, South Korea, Japan, China, Singapore, etc. which were not caused by imported cases [1, 136, 137]. To finally overcome the pandemic, it is very important to identify all the infected, and this is very difficult due to the large number of asymptomatic patients. It is possible that the experience of Slovakia, which tested once more 24% of the population on November 7 and found 0.63% of those infected [138], will be useful for other countries.

Adequate modeling is further complicated by the fact that we do not know when the number of reported cases is approaching the actual number. Therefore, it remains to repeat the calculations using more and more recent data sets and compare the resulting curves with the subsequent dynamics of epidemics, as was done in the cases of Ukraine, Italy, and Austria (see Chap. 6). Such an approach ensured a fairly high accuracy of the first pandemic wave forecasts for Austria, Italy, Spain, France, and Germany (see Table 7.4).

The second reason for the limited accuracy of long-term forecasts is the constant changes in the conditions of the pandemic (changing quarantine measures, social behavior, virulence of the pathogen, etc.). Therefore, the prediction made using statistics for a certain period T_c is no longer suitable for other periods of time. To illustrate this fact, we take two predictions: for Ukraine (wave 4, see Table 11.1) and for the world (wave 5, Table 12.2). Table 13.1 illustrates a very high accuracy of short-time predictions. It can be seen that after one week of observations (after the last day of the periods taken for calculations T_c) the real accumulated number of cases exceeds the calculated ones by 0.54% and 1.35% for Ukraine and the world,

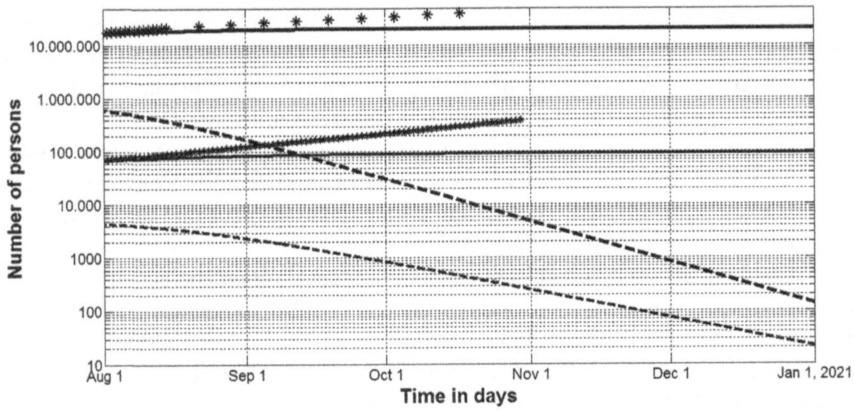

Fig. 13.1 Long-term predictions for Ukraine (fourth wave, Table 11.1) and the world (fifth wave, Table 12.2). SIR curves (lines) and accumulated number of cases ("stars") versus time. Bold lines and large markers correspond to the world dynamics. Numbers of infected and spreading I (dashed lines) and victims (accumulated number of confirmed cases) $V = I + R$ (solid lines)

Table 13.1 Comparison of the theoretical results with the real number of cases after one and two weeks of observations

Characteristics	Ukraine, fourth wave, July 23, 2020	Ukraine, fourth wave, July 30, 2020	World, fifth wave, July 25, 2020	World, fifth wave, August 1, 2020
Calculated number of cases $V = I + R$	62,486	67,571	15,571,846	16,909,569
Confirmed number of cases V_j, WHO reports, [1]	62,823	69,884	15,785,641	17,660,523
Discrepancy (%)	0.54	3.31	1.35	4.25

respectively. After two weeks of observations, the discrepancies are higher (3.31% for Ukraine and 4.25% for the world).

But over time, these discrepancies have increased (compare solid lines and markers in Fig. 13.1) and as of the end of October 2020, the real number of cases was 2–3 times higher than the predicted saturation levels $V_{i\infty}$. Figure 13.1 shows that corresponding long-term predictions for Ukraine and the world for the period August–December 2020 were very optimistic. The new pandemic waves occurred due to the quarantine easing and changes in social distancing. Some of them were already detected and calculated in Chaps. 10 and 11. Gradual weakening of quarantines may not cause pronounced new waves. In this case, the parameters of the pandemic dynamics will change continuously, which makes the proposed approach unsuitable. In any case, the global and Ukrainian dynamics must be updated with the use of new data sets. Because of this, there is a need for simple methods to assess such important characteristics of epidemics as their duration and the final number of cases. We will try to develop such approaches using the experience gained in calculating the epidemic waves in Ukraine and the whole world.

First, we can use the values, which are known at the beginning t_s^* of a new epidemic wave: the accumulated number of cases V_s and daily number of new cases d_s. Since these characteristics are random, we will use the smoothed values according to formulas (8.1) and (8.2):

$$V_s = \bar{V}_i\big|_{t=t_s^*} = \frac{1}{7}\sum_{j=i-3}^{j=i+3} V_j, \tag{13.1}$$

$$d_s = \frac{d\bar{V}}{dt}\bigg|_{t=t_s^*} \approx \frac{1}{2}\left(\bar{V}_{i+1} - \bar{V}_{i-1}\right) \tag{13.2}$$

Index i in formulas (13.1) and (13.2) corresponds to the moment of time t_s^*. Since the calculations need V_j values for some previous moments of time (see (13.1) and (13.2)), values t_s^* shown in Table 13.2 are larger than time moments t_i^* corresponding to the beginning of the ith epidemic wave. The results of calculations with the use of Eqs. (13.1) and (13.2) are presented in Table 13.2.

Table 13.2 Characteristics of the COVID-19 pandemic waves in Ukraine and the world

Country or region, number of wave (prediction)	Day used for calculations t_s^*	V_s, Eq. (13.1)	d_s, Eq. (13.2)	d_s/V_s	$V_{i\infty}/V_s$	T_s (days), Eq. (13.3)	D_p, Eq. (13.4)
Ukraine, wave 1 (prediction 8)	May 7	14,194	474.5	0.0334	2.00	156.9	5.25
Ukraine, wave 2	May 21	20,110	381.2	0.0190	3.03	409.9	7.77
Ukraine, wave 3	June 13	31,143	703.3	0.0226	2.28	193.5	4.37
Ukraine, wave 4	July 7	50,505	737.6	0.0146	1.84	252.7	3.69
Ukraine, wave 5	July 24	63,811	929.4	0.0146	2.02	195	2.84
Ukraine, wave 6	August 13	87,939	1,584.6	0.0180	2.79	237.5	4.28
Ukraine, wave 7	October 5	235,459	4,773.0	0.0203	1.69	173	3.51
World, wave 1 (prediction 2)	May 2	3,346,376	8,2258.1	0.0246	2.08	312.5	7.68
World, wave 2	May 16	4,525,359	90,255.6	0.0120	4.51	632.5	12.61
World, wave 3	June 1	6,177,809	119,995.4	0.0194	2.50	519.3	10.09
World, wave 4	June 17	8,241,160	146,738.7	0.0178	2.77	451	8.03
World, wave 5	July 9	12,112,797	207,501.7	0.0171	1.72	256.2	4.39

We have also calculated the duration of the epidemic wave T_s in days after moment of time t_s^* according to the equation:

$$T_s = t_{if} - t_s^* \tag{13.3}$$

and corresponding dimensionless characteristics:

$$D_p = \frac{T_s d_s}{V_s} \tag{13.4}$$

These values and characteristics d_s/V_s and $V_{i\infty}/V_s$ are also presented in Table 13.2.

Figure 13.2 illustrates the results of calculations versus time. The values of characteristics are shown at the moments of time t_s^* for Ukraine (small markers) and the world (large markers). It can be seen that the values of parameters: D_p ("squares"), $V_{i\infty}/V_s$ ("crosses") and d_s/V_s ("stars") are rather close for the epidemic waves which occurred after July 1, 2020. The same picture we can see in Fig. 13.3 showing the values of the dimensionless final size of the ith wave $V_{i\infty}/V_s$ ("circles") and duration after prediction T_s (Eq. (13.3), "stars") versus dimensionless parameter D_p calculated for pandemic waves in Ukraine (small markers) and the world (large markers).

If we are interested in long-term forecasts of the pandemic dynamics after September 2020, it is worth taking into account the values of the parameters obtained after July 1, 2020. Then, Table 13.2 and Figs. 13.2 and 13.3 allow us to determine the limits for the dimensionless parameters $V_{i\infty}/V_s$ and D_p and to

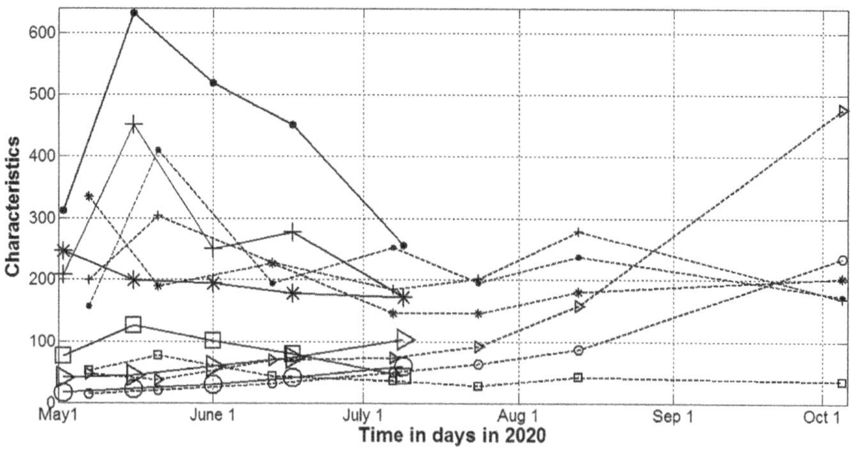

Fig. 13.2 Characteristics of the COVID-19 pandemic waves in Ukraine (small markers) and the world (large markers) versus time: "circles" represent the averaged accumulated number of cases V_s (Eq. (13.1), for Ukraine these values are divided by 1000, for the world by 200,000); "triangles" show the averaged daily number of new cases d_s (Eq. (13.2), for Ukraine these values are divided by 10, for the world by 2000); "squares"—dimensionless parameter D_p (Eq. (13.4)); "crosses" represent dimensionless final size of the ith wave $V_{i\infty}/V_s$ (multiplied by 100); "stars"—d_s/V_s (multiplied by 10,000); "dots"—duration after prediction T_s (Eq. (13.3))

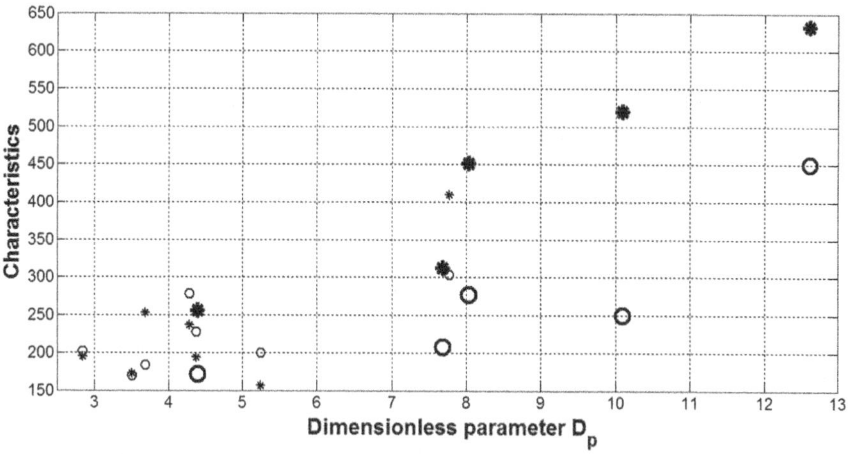

Fig. 13.3 Characteristics of the COVID-19 pandemic waves in Ukraine (small markers) and the world (large markers) versus dimensionless parameter D_p: "circles" represent the dimensionless final size of the ith wave $V_{i\infty}/V_s$ (multiplied by 100); "stars"—duration after prediction T_s (Eq. (13.3))

estimate the lower and upper limits for the final size of a new wave and its duration after prediction T_s:

$$1.69 \cdot V_s \leq V_{i\infty} \leq 2.79 \cdot V_s \tag{13.5}$$

$$2.84\frac{V_s}{d_s} \leq T_s \leq 4.39\frac{V_s}{d_s} \tag{13.6}$$

Estimations (13.5) and (13.6) were applied for new epidemic waves in different countries and regions. The results of calculations are shown in Table 13.3. To calculate the values V_s and d_s, the weekly WHO data for accumulated number of cases w_i were used [1], according to the formulas:

$$V_s \approx \frac{w_i + w_{i-1}}{2}, \tag{13.7}$$

$$d_s \approx \frac{w_i - w_{i-1}}{7} \tag{13.8}$$

The days corresponding w_i and w_{i-1} were chosen with the use of Figs. 8.1–8.11, 8.13, 8.14 which allows us to detect the beginning of new pandemic waves (sharp increase in the second derivative of the accumulated number of cases shown by "stars"). In many countries, we can see the beginning of the new wave in early October 2020 that is why the days October 4 and October 11 were used in (13.7) and (13.8). In some countries, the new waves have started earlier. The corresponding days for South Korea, Spain, Moldova, and Ukraine are shown in Table 13.3. To have a more recent forecast for Kyiv, we have used the period October 12–17 (corresponding to the beginning of the new wave seen in Fig. 8.14), the daily data set from Table 8.2 and formulae (13.1) and (13.2). The predicted lower and upper limits of the final sizes $V_{i\infty}$ were compared with the real figures for accumulated numbers of cases V_j as of October 31, 2020 [1] (November 9, 2020, was used for Kyiv [95, 96]).

The results of calculations are presented in Table 13.3. It can be seen that V_j values exceed the upper limit of $V_{i\infty}$ only in the case of Spain, but the period of observations was 9 weeks for this country (in comparison with 3–5 weeks for other countries and regions). It must be noted that V_j number in South Korea did not exceed even the lower limit of $V_{i\infty}$ after 10 weeks of observations. The results of the calculations allow us to conclude that formula (13.5) yields adequate forecasts of the number of cases for a fairly long period, provided that the nature of the pandemic will be as at the beginning of the corresponding wave. Some European countries have tightened quarantines in November 2020, so we hope that the numbers of cases will not exceed the upper limits shown in Table 13.3 for a long time. Otherwise, new estimates should be made using the simulations proposed in Chaps. 9 and 10 or formulas (13.5) and (13.6).

The pandemic duration projections presented in Table 13.3 are unfortunately not optimistic. None of the countries will overcome the current wave of the epidemic in 2020. In Germany and Italy, it can last until the autumn of 2021, and in USA until

Table 13.3 Predictions for the COVID-19 pandemic waves in different countries and regions

Country or region	Period used for calculations	V_s, Eq. (13.7) or (13.1)	d_s, Eq. (13.8) or (13.2)	$V_{i\infty}$ lower limit, Eq. (13.5)	V_j as of October 31, 2020 (November 9 for Kyiv)	$V_{i\infty}$ upper limit, Eq. (13.5)	T_s (days) lower limit, Eq. (13.6)	T_s (days) upper limit, Eq. (13.6)
World	October 4–11	35,957,100	329,360	60,767,499	45,968,799	100,320,309	310	479
South Korea	August 15–23	16,358.5	297.3	27,646	26,635	45,640	156	241
Ukraine	September 20–27	187,156	3,279.4	316,294	395,440	522,165	162	250
Sweden	September 13–20	87,371	247.4	147,657	124,355	243,765	1,003	1,550
USA	October 4–11	7,419,991	46,788	12,539,785	8,952,086	20,701,774	450	696
Germany	October 4–11	311,050.5	3,375.3	525,675	532,930	867,831	261	405
UK	October 4–11	535,434.5	15,832	904,884	1,011,664	1,493,862	96	148
Italy	October 4–11	336,122.5	3,820.4	568,047	679,430	937,782	249	386
Spain	August 23–30	412,670	7,604.6	697,412	1,185,678	1,151,349	154	238
France	October 4–11	636,035.5	15,809	1,074,900	1,331,808	1,774,539	114	177
Austria	October 4–11	51,352	952.3	86,785	106,584	143,272	153	237
Moldova	September 20–27	48,435	599.7	81,855	76,040	135,134	229	355
Kyiv	October 11–17	29,078	395.7	49,142	45,667	81,128	209	323

the end of 2022. Such a long-lasting pandemic is in line with the forecast for Africa [104], according to which new COVID-19 cases will cease to appear only after September 2021. Very surprising are the results for Sweden. Very long duration of the epidemic wave (3–4 year) can be connected with the absence of lockdown in this country. Maybe formula (13.6) developed for the cases with lockdown is not applicable for Sweden. The final size of the recent pandemic wave can exceed 100 million cases in the world.

The forecasts given in Table 13.3 can usually be applied for a limited period of time due to constant changes in the pandemic dynamics. For example, we have used the period T_c: October 1–14 for SIR simulations of the seventh epidemic wave in Ukraine (see Table 10.1). Obtained value $V_{7\infty} = 398317$ and the final day of this wave: March 27, 2021, are in good agreement with the inequalities (13.5), (13.6) and data for Ukraine given in Table 13.3.

In Chap. 10, we have detected new COVID-19 epidemic waves in Ukraine caused by the local elections and the presidential poll, which took place across the country on October 25, 2020. The apparent increase in contacts even before they took place clearly accelerated the growth in the number of cases. Since the largest value of the second derivative (Eq. (8.3)) corresponded to October 31 (see Fig. 10.1), we chose this day as the beginning of the eighth wave and calculated the values V_s and daily number of new cases d_s with the use of (13.1) and (13.2) and data set presented in Table 8.2. The resulting figures $V_s = 395139$ and $d_s = 8399.4$ allow us to calculate lower and upper limits for the final size $V_{i\infty}$ and epidemic duration T_s (in days) after October 31, 2020, with the use of estimations (13.5) and (13.6):

$$667785 \leq V_{8\infty} \leq 1102438 \tag{13.9}$$

$$133.6 \leq T_s \leq 206.5 \tag{13.10}$$

Comparison of inequality (13.9) with the corresponding values in Table 13.3 shows that the elections and the presidential poll approximately doubled the expected number of cases. The final number of accumulated cases in Ukraine can be greater than one million. To compare the predictions for the final day of the epidemic in Ukraine, it is necessary to subtract 37 days from the values given in Table 13.3 (to take into account the difference in the values of t_s^* corresponding to September 24 and October 31). After doing this procedure and comparing the results with inequality (13.10), we see that estimates of the duration of the eighth epidemic wave have not changed (they even became slightly smaller than for the previous wave). Simple calculations using (13.10) show that the end of the eighth epidemic wave can be expected from March 14 to May 26, 2021. All the calculations given in this book gave only underestimated values in relation to the number of cases. We hope that effective quarantine measures and successful vaccination will be able to reverse these trends.

Chapter 14
Conclusions

The accuracy of any mathematical model is limited. Unfortunately, the correct mathematical simulation of the COVID-19 pandemic is especially difficult for at least two reasons. First, data on the number of cases were clearly incomplete immediately after outbreaks; there were quite long hidden periods before the first case in a country or region was confirmed. Large discrepancies between the registered and actual number of cases occurred even for the later periods of the COVID-19 pandemic. The second reason for the limited accuracy of mathematical modelling is the constant changes in the conditions of the pandemic (quarantine measures, social behavior, virulence of the pathogen, etc.). Therefore, the prediction made using statistics for a certain period is no longer suitable for other periods of time.

The accuracy of mathematical simulation could be improved if we could use the numbers of cases corresponding to the moments of time when persons got sick. Unfortunately, the official WHO data sets deal with the days when the corresponding numbers of cases were laboratory confirmed. The delay can be different in different countries and rather long. Nevertheless, the results of our research have shown that mathematical models can adequately describe and predict the dynamics of both the initial stages of the pandemic and its subsequent waves.

Exponential growth of the number of cases typical for every epidemic was revealed also in COVID-19 dynamics for some countries and regions in March–April 2020. Corresponding duplication rates and basic reproduction numbers were calculated. Fortunately, after late March the dynamics in many countries and in the world showed signs of stabilization (exponential growth stopped). After the end of the summer of 2020 new powerful waves began (especially in European countries and the United States).

In the book, some methods of comparison of the epidemic dynamics in different countries were proposed. This allowed us to use information about regions where the epidemic started earlier to predict its dynamics immediately after the outbreaks in other countries. In particular, it was calculated that the fist epidemic wave in Austria would stop with the number of cases similar to the corresponding figure in

I. Nesteruk, *COVID-19 Pandemic Dynamics*,
https://doi.org/10.1007/978-981-33-6416-5_14

the Republic of Korea. Already in early March 2020 it was predicted that the number of cases in Italy will be higher than in China.

The numbers of cases versus the epidemic duration (number of days after the first COVID-19 case was confirmed) were compared for Ukraine and its neighboring countries. An attempt to assess the situation only by the number of cases on a fixed day after epidemic outbreak led to some wrong predictions due to the problems with identification of infected persons. Therefore, other factors (including possible exponential growth) must also be taken into account.

The classical SIR model was applied to simulate the first epidemic waves in many countries and regions. This model contains four parameters and three differential equations for the evolution of the number of susceptible persons—S; infected, spreading the infection—I and removed persons—R, which is the sum of isolated, immunized, and deceased persons. The unknown values of parameters were successfully estimated using the known exact solution of the differential equations and observations of the accumulated number of COVID-19 cases. The adequate predictions for the expected numbers of cases and the durations of the first wave have been obtained. The difference between the predicted saturation levels and the real number of cases on final days of the first waves is 4–17% for China, Germany, the Republic of Korea, Austria, Italy, Spain, and France. Worse accuracy for USA, the UK, the Republic of Moldova, Ukraine, the city of Kyiv and the whole world can be explained by the fact that the next waves of the epidemic began before the end of the first due to lack of proper quarantine, mass violations of social distance and problems with testing.

The average times of spreading the infection, reproduction numbers, probability of meeting an infected person, and other characteristics of the first epidemic waves were calculated and compared for different countries and WHO regions. The results of SIR simulations demonstrate that the pandemic probably began in August 2019. Some approximate formulas linking the final sizes and infection rates with the volume of population were developed. Relatively small number of people who spread COVID-19 can cause a fairly long course of the epidemic. In particular, it was shown that 20 infected persons in Ukraine can cause new cases for another two and a half months. To finally overcome the pandemic, it is very important to identify all the infected, and this is very difficult due to the large number of asymptomatic patients. Probably, experience of Slovakia, which tested the majority of population in two days, will be useful for other countries.

Constant changes in the COVID-19 pandemic conditions (i.e., in the peculiarities of quarantine and its violation, in situations with testing and isolation of patients, in coronavirus activity due to its mutations, etc.) cause changes in the values of parameters of the mathematical models and lead to new pandemic waves. In particular, in October 2020 we observed a sharp increase in the daily number of new cases in many European countries. To identify these changes, a simple method was proposed based on the numerical differentiation of smoothed dependences of the accumulated number of cases. This approach was successfully applied to detect the epidemic waves in USA, Germany, the UK, the Republic of Korea, Italy, Austria, Spain, France, Sweden, the Republic of Moldova, Ukraine, the city of Kyiv and the whole world. The results show that smoothing the dependence of the accumulated number of cases and its differentiation can provide fairly accurate and useful information about the course of the epidemic, identify important changes in its dynamics and provide timely recommendations for quarantine measures or control of social distancing.

To simulate different pandemic waves (periods with more or less constant values of its dynamics parameters), a new general SIR model and its exact solution have been proposed. Two procedures of its parameters identification were developed and successfully applied to calculate the characteristics of several pandemic waves in Ukraine and the world.

Due to the incomplete number of detected cases and permanent changes in pandemic dynamics, SIR simulations must be updated using new data sets in order to have good accuracy of predictions. The number of observations has to be large enough, since all the conclusions are based on corresponding data sets. To avoid complicated calculations and to obtain some predictions immediately after the beginning of a new wave, simple estimations for the final size of a pandemic wave and its duration were proposed and applied for USA, Germany, the UK, the Republic of Korea, Italy, Austria, Spain, France, the Republic of Moldova, Sweden, Ukraine, the city of Kyiv and the whole world.

The pandemic duration projections are unfortunately not optimistic. None of the countries would overcome the current wave of the epidemic in 2020. In Germany and Italy, it can last until the autumn of 2021, in the United States until the end of 2022. The final size of the recent pandemic wave can exceed 100 millions victims in the world. All the calculations given in this book gave only underestimated values in relation to the number of cases. We hope that effective quarantine measures and successful vaccination will be able to reverse this trend.

Very long duration of the pandemic requires correction of our behavior, we cannot live as before it occurred. Decreased feelings of insecurity and non-compliance with social distancing may further increase the pandemic duration and the number of the coronavirus victims. Total closure of settlements or regions can be recommended only in the event of a sharp increase in the number of cases. There are many things that can be done without loss to the economy and our daily lives:

1. Minimize the number of contacts and trips, do not visit crowded places. Work and study remotely where possible.
2. Refrain from shaking hands and kisses during meetings. Use masks in transport and crowded areas.
3. If you (or others) have any suspicious symptoms, do your best to avoid the spread of the infection.

References

1. World Health Organization. Coronavirus disease (COVID-2019) situation reports. https://www.who.int/emergencies/diseases/novel-coronavirus-2019/situation-reports/. Retrieved 3 Oct 2020
2. https://en.wikipedia.org/wiki/Timeline_of_the_2019%E2%80%9320_Wuhan_coronavirus_outbreak. Retrieved 15 Feb 2020
3. https://www.theguardian.com/world/2020/feb/23/italy-draconian-measures-effort-halt-coronavirus-outbreak-spread
4. http://www.salute.gov.it/portale/nuovocoronavirus/dettaglioContenutiNuovoCoronavirus.jsp?lingua=italiano&id=5351&area=nuovoCoronavirus&menu=vuoto
5. https://www.ecdc.europa.eu/en/covid-19/interactive-map
6. Li Q, Guan X, Wu P et al (2020) Early transmission dynamics in Wuhan, China, of novel coronavirus–infected pneumonia. The New England J med https://doi.org/10.1056/nejmoa2001316
7. Wu JT, Leung K, Leung GM (2020) Nowcasting and forecasting the potential domestic and international spread of the 2019-nCoV outbreak originating in Wuhan, China: a modelling study. Lancet. https://doi.org/10.1016/s0140-6736(20)30260-9
8. Zhao S, Lin Q, Ran J, Musa SS, Yang G, Wang W, Lou Y, Gao D et al (2020) Preliminary estimation of the basic reproduction number of novel coronavirus (2019-nCoV) in China, from 2019 to 2020: a data-driven analysis in the early phase of the outbreak. Int J Infect Dis. https://doi.org/10.1016/j.ijid.2020.01.050
9. Byass P (2020) Eco-epidemiological assessment of the COVID-19 epidemic in China, January-Februry 2020. medRxiv 2020. https://doi.org/10.1101/2020.03.29.20046565
10. Zhao S, Lin Q, Ran J et al (2020) Preliminary estimation of the basic reproduction number of novel coronavirus (2019-nCoV) in China, from 2019 to 2020: a data-driven analysis in the early phase of the outbreak. Int J Infect Dis 92:214–217. https://doi.org/10.1016/j.ijid.2020.01.050
11. Tang B, Bragazzi NL, Li Q et al (2020) An updated estimation of the risk of transmission of the novel coronavirus (2019-nCov). Infect Dis Model 5:248–255. https://doi.org/10.1016/j.idm.2020.02.001
12. Report of the WHO–China joint mission on coronavirus disease 2019 (COVID-19). https://www.who.int/docs/default-source/coronaviruse/who-china-joint-mission-on-covid-19-final-report.pdf

I. Nesteruk, *COVID-19 Pandemic Dynamics*,
https://doi.org/10.1007/978-981-33-6416-5

13. Ferguson NM et al (2020) Impact of non-pharmaceutical interventions (NPIs) to reduce COVID-19 mortality and health care demand. Preprint, Imperial College COVID-19 Response Team
14. Liu Y, Gayle AA, Wilder-Smith A, Rocklöv J (2020) The reproductive number of COVID-19 is higher compared to SARS coronavirus. J Travel Medicine, 27(2), taaa021, https://doi.org/10.1093/jtm/taaa021
15. Li Y, Liang M, Yin X et al (2020) COVID-19 epidemic outside China: 34 founders and exponential growth. [Preprint.] medRxiv. https://doi.org/10.1101/2020.03.01.20029819
16. Kucharski AJ et al (2020) Early dynamics of transmission and control of COVID-19: a mathematical modelling study. Lancet Infect Dis. https://doi.org/10.1016/S1473-3099(20)30144-4
17. Batista M (2020) Estimation of the final size of the COVID-19 epidemic. [Preprint.] medRxiv. https://doi.org/10.1101/2020.02.16.20023606
18. Hong HG, Li Y (2020) Estimation of time-varying reproduction numbers underlying epidemiological processes: a new statistical tool for the COVID-19 pandemic. PLoS ONE 15(7):e0236464. https://doi.org/10.1371/journal.pone.0236464
19. Dehning J et al (2020) Inferring COVID-19 spreading rates and potential change points for case number forecasts. Preprint, ArXiv:2004.01105
20. Chen Y, Cheng J, Jiang Y, Liu K (2020) A time delay dynamical model for outbreak of 2019-nCoV and the parameter identification. ArXiv:2002.00418
21. Peng L, Yang W, Zhang D, Zhuge C, Hong L (2020) Epidemic analysis of COVID-19 in China by dynamical modeling. ArXiv:2002.06563
22. Chang SL, Harding N, Zachreson C, Cliff OM, Prokopenko M (2020) Modelling transmission and control of the COVID-19 pandemic in Australia. ArXiv:2003.10218
23. Pellis L, Scarabel F, Stage HB, Overton CE, Chappell LH, Lythgoe KA et al (2020) Challenges in control of Covid-19: short doubling time and long delay to effect of interventions. ArXiv:2004.00117
24. Zhou T, Liu Q, Yang Z, Liao J, Yang K, Bai W et al (2020) Preliminary prediction of the basic reproduction number of the Wuhan novel coronavirus 2019-nCoV. J Evid-Based Med. https://doi.org/10.1111/jebm.12376PMID:3204881523
25. Maier BF, Brockmann D (2020) Effective containment explains sub-exponential growth in confirmed cases of recent COVID-19 out break in mainland China. ArXiv:2002.07572
26. Song PX, Wang L, Zhou Y, He J, Zhu B, Wang F et al (2020) An epidemiological forecast model and software assessing interventions on COVID-19 epidemic in China. MedRxiv
27. Chinazzi M, Davis JT, Ajelli M, Gioannini C, Litvinova M, Merler S et al (2020) The effect of travel restrictions on the spread of the 2019 novel coronavirus (COVID-19) outbreak. Science 368(6489):395–400. https://doi.org/10.1126/science.aba9757 PMID:32144116
28. Zhang Y, Jiang B, Yuan J, Tao Y (2020) The impact of social distancing and epicenter lockdown on the COVID-19 epidemic in mainland China: a data-driven SEIQR model study. MedRxiv
29. Benlagha N (2020) Modeling the declared new cases of COVID-19 trend using advanced statistical approaches. [Preprint.] ResearchGate. March 2020. https://doi.org/10.6084/m9.figshare.12052638
30. Udomsamuthirun P et al (2020) The reproductive index from SEIR model of Covid-19 epidemic in Asean. [Preprint.] MEDRXIV. https://doi.org/10.1101/2020.04.24.20078287
31. Pereira IG et al (2020) Forecasting Covid-19 dynamics in Brazil: a data driven approach. Int J Environ Res Public Health 17(14):5115. https://doi.org/10.3390/ijerph17145115
32. Linka K, Peirlinck M, Kuhl E (2020) The reproduction number of COVID-19 and its correlation with public health interventions. [Preprint.] MEDRXIV. https://doi.org/10.1101/2020.05.01.20088047
33. Draper NR, Smith H (1998) Applied regression analysis (3rd edn) John Wiley
34. https://onlinepubs.trb.org/onlinepubs/nchrp/cd-22/manual/v2appendixc.pdf

35. Nesteruk I (2017) Statistics based models for the dynamics of Chernivtsi children disease, Naukovi Visti NTUU KPI 5:26–34. https://doi.org/10.20535/1810-0546.2017.5.108577
36. Gazzola M, Argentina M, Mahadevan L (2014) Scaling macroscopic aquatic locomotion. Nat Phys 10:758–761
37. Nesteruk I (2019) Maximal speed of underwater locomotion. Innov Biosyst Bioeng 3 (3):152–167. https://doi.org/10.20535/ibb.2019.3.3.177976
38. Nesteruk I (2020) Stabilization of the coronavirus pandemic in Italy and global prospects. [Preprint.] MEDRXIV. https://medrxiv.org/cgi/content/short/2020.03.28.20045898v1
39. Nesteruk I, Kydybyn I, Demelmair G (2020) Coronavirus pandemic dynamics in March. 2020. What can we expect in April?. [Preprint.] ResearchGate. https://doi.org/10.13140/rg.2.2.36305.28009
40. Nesteruk I, Kydybyn I, Demelmair G (2020) Corona pandemic global stabilization? [Preprint.] ResearchGate. https://doi.org/10.13140/rg.2.2.21570.32961. https://www.researchgate.net/publication/340917471_Corona_pandemic_global_stabilization
41. Nesteruk I, Kydybyn I, Demelmair G (2020) Global stabilization trends Of Covid-19 pandemic. KPI Sci News 2:55–62. https://doi.org/10.20535/kpi-sn.2020.2.205124
42. Nesteruk I (2020) How can we estimate the dangers of the coronavirus epidemic in Europe? [Preprint.] ResearchGate. https://doi.org/10.13140/rg.2.2.32072.06402
43. Nesteruk I (2020) Coronavirus epidemic outbreak in Europe. Comparison with the dynamics in mainland China. KPI Sci News 1:15–17. https://doi.org/10.20535/kpi-sn.2020.1.198043
44. Distante C, Gadelha Pereira I (2020) Garcia Goncalves LM, Piscitelli P, Miani A (2020) Forecasting Covid-19 outbreak progression in Italian regions: a model based on neural network training from Chinese data. medRxiv
45. Hamzah F, Binti A, Lau C, Nazri H, Ligot DV, Lee G, Tan CL (2020) Corona tracker: worldwide COVID-19 outbreak data analysis and prediction. Bull World Health Organ 1:32
46. Fanelli D, Piazza F (2020) Analysis and forecast of COVID-19 spreading in China, Italy France. Chaos Solitons Fractals 134:109761
47. Webb GF, Magal P, Liu Z, Seydi O (2020) A model to predict COVID-19 epidemics with applications to South Korea, Italy, and Spain. medRxiv
48. Bastos SB (2020) In: Cajueiro DO (ed) Modeling and forecasting the early evolution of the Covid-19 pandemic in Brazil. arXiv:2003.14288
49. Nesteruk I (2020) Comparison of the coronavirus pandemic dynamics in Europe, USA and South Korea. [Preprint.] MEDRXIV. https://medrxiv.org/cgi/content/short/2020.03.18.20038133v1
50. Nesteruk I (2020) Comparison of the coronavirus epidemic dynamics in Italy and mainland China [Preprint.] ResearchGate. https://doi.org/10.13140/rg.2.2.19152.87049
51. Nesteruk I (2020) Comparison of the coronavirus pandemic dynamics in Ukraine and neighboring countries. [Preprint.] ResearchGate. https://doi.org/10.13140/rg.2.2.31170.53448
52. https://www.who.int/docs/default-source/coronaviruse/situation-reports/20200928-weekly-epi-update.pdf?sfvrsn=9e354665_6. Retrieved 1 Oct 2020
53. Kermack WD, McKendrick AG (1927) A contribution to the mathematical theory of epidemics. J Roy Stat Soc Ser A 115:700–721
54. Murray JD (2002) Mathematical biology I/II. Springer, New York
55. Bailey NTJ (1957) The mathematical theory of epidemics. Griffin Book Co
56. Langemann D, Nesteruk I, Prestin J (2016) Comparison of mathematical models for the dynamics of the Chernivtsi children disease. Math Comput Simul 123:68–79
57. Grant A (2020) Dynamics of COVID-19 epidemics: SEIR models underestimate peak infection rates and overestimate epidemic duration. medRxiv
58. Loli Piccolomiini E, Zama F (2020) Monitoring Italian COVID-19 spread by an adaptive SEIRD model. medRxiv

59. Baerwolff GK (2020) A contribution to the mathematical modeling of the Corona/COVID-19 Pandemic. medRxiv
60. Distante C, Piscitelli P, Miani A (2020) Covid-19 outbreak progression in Italian regions: approaching the peak by the end of march in northern Italy and first week of april in southern Italy. Int J Environ Res Public Health 17:3025
61. Wang L, Li J, Guo S, Xie N, Yao L, Cao Y, Day SW, Howard SC, Graff JC, Gu T et al (2020) Real-time estimation and prediction of mortality caused by COVID-19 with patient information based algorithm. Sci Total Environ 727:138394
62. te Vrugt M, Bickmann J, Wittkowski R (2020) Effects of social distancing and isolation on epidemic spreading: a dynamical density functional theory model. arXiv:2003.13967
63. Yang Z, Zeng Z, Wang K, Wong SS, Liang W, Zanin M, Liu P, Cao X, Gao Z, Mai Z et al (2020) Modified SEIR and AI prediction of the epidemics trend of COVID-19 in China under public health interventions. J Thorac Dis 12:165
64. Roda WC, Varughese MB, Han D, Li MY (2020) Why is it difficult to accurately predict the COVID-19 epidemic? Infect Dis Model 5:271–281
65. Otunuga OM, Ogunsolu MO (2020) Qualitative analysis of a stochastic SEITR epidemic model with multiple stages of infection and treatment. Infect Dis Model 5:61–90
66. Nesteruk I (2020) Statistics based predictions of coronavirus 2019-nCoV spreading in mainland China, MedRxiv Posted. https://doi.org/10.1101/2020.02.12.20021931
67. Nesteruk I (2020) Statistics-based predictions of coronavirus epidemic spreading in mainland China. Inntive Biosyst Bioeng 4(1):13–18. https://doi.org/10.20535/ibb.2020.4.1.195074
68. Nesteruk I (2020) Verification of coronavirus epidemic predictions for mainland China. [Preprint.] ResearchGate. https://doi.org/10.13140/rg.2.2.17231.43681. https://www.researchgate.net/publication/339285703_Verification_of_coronavirus_epidemic_predictions_for_mainland_China?channel=doi&linkId=5e47d59d458515072d9e26d8&showFulltext=true
69. Nesteruk I (2020) Characteristics of coronavirus epidemic in mainland China estimated with the use of official data available after February 12, 2020. [Preprint.] ResearchGate. https://doi.org/10.13140/rg.2.2.19667.32804
70. Nesteruk I (2020) Estimations of the coronavirus epidemic dynamics in South Korea with the use of SIR model. [Preprint.] ResearchGate. https://doi.org/10.13140/rg.2.2.15489.40807
71. Nesteruk I (2020) Long-term predictions for COVID-19 pandemic dynamics in Ukraine, Austria and Italy. [Preprint.] MEDRXIV https://www.medrxiv.org/content/10.1101/2020.04.08.20058123v1
72. Nesteruk I (2020) Як довго українці сидітимуть на карантині? How long will the Ukrainians stay in quarantine? (in Ukrainian) [Preprint.] ResearchGate. https://doi.org/10.13140/rg.2.2.15732.71046
73. Nesteruk I (2020) Динаміка COVID-19 епідемії в Україні та Києві після покращання тестування. COVID-19 epidemic dynamics in Ukraine and Kyiv after testing has improved (In Ukrainian). [Preprint.] ResearchGate. https://doi.org/10.13140/rg.2.2.36705.86885 https://www.researchgate.net/publication/341043366_Dinamika_COVID-19_epidemii_v_Ukraini_ta_Kievi_pisla_pokrasanna_testuvanna_COVID-19_epidemic_dynamics_in_Ukraine_and_Kyiv_after_testing_has_improved
74. Nesteruk I (2020) SIR-simulation of Corona pandemic dynamics in Europe. [Preprint.] MEDRXIV, https://medrxiv.org/cgi/content/short/2020.04.22.20075135v1
75. Nesteruk I (2020) Hidden periods, duration and final size of COVID-19 pandemic. Preprint. MedRxiv. https://medrxiv.org/cgi/content/short/2020.05.10.20097147v1
76. Nesteruk I (2020) Статистика пандемії COVID-19 в Україні та світі COVID-19. Pandemic statistics in Ukraine and world. (In Ukrainian). ResearchGate. https://doi.org/10.13140/rg.2.2.30999.39842. https://www.researchgate.net/publication/341542317_Statistika_pandemii_COVID-19_v_Ukraini_ta_sviti_COVID-19_pandemic_statistics_in_Ukraine_and_world

77. Nesteruk I (2020) Simulations and predictions of COVID-19 pandemic with the use of SIR model. Innov Biosyst Bioeng 4(2):110–121. https://doi.org/10.20535/ibb.2020.4.2.204274. http://ibb.kpi.ua/article/view/204274
78. Nesteruk I (2020) Coronasummer in Ukraine and Austria. [Preprint.] ResearchGate. https://doi.org/10.13140/rg.2.2.32738.56002
79. https://en.wikipedia.org/wiki/Basic_reproduction_number
80. Kyiv City Administartion. https://health.kievcity.gov.ua/
81. Nesteruk I (2020) Statistics based predictions of coronavirus 2019-nCoV spreading in mainland China, [Preprint.] ResearchGate. Posted https://doi.org/10.13140/rg.2.2.32953.52322
82. https://www.who.int/docs/default-source/coronaviruse/situation-reports/20200824-weekly-epi-update.pdf?sfvrsn=806986d1_4
83. Nesteruk I (2020) SIR-simulation of Corona pandemic dynamics in Europe. [Preprint.] Research Gate. https://doi.org/10.13140/rg.2.2.13542.16969
84. https://www.scmp.com/news/china/society/article/3076334/coronavirus-strange-pneumonia-seen-lombardy-november-leading
85. Lescure F-X et al (2020) Clinical and virological data of the first cases of COVID-19 in Europe: a case series. www.thelancet.com/infection. https://doi.org/10.1016/S1473-3099(20)30200-0
86. https://www.who.int/docs/default-source/coronaviruse/situation-reports/20200928-weekly-epi-update.pdf?sfvrsn=9e354665_6
87. Coronavirus: trump considering quarantining New York. BBC News. https://www.bbc.com/news/world-us-canada-52079121
88. Weinberger DM et al (2020) Estimating the early death toll of COVID-19 in the United States. [Preprint.] MEDRXIV, https://doi.org/10.1101/2020.04.15.2006643
89. Nesteruk I (2020) Waves of COVID-19 pandemic. Detection and SIR simulations. MedRxiv. https://doi.org/10.1101/2020.08.03.20167098. https://www.medrxiv.org/content/10.1101/2020.08.03.20167098v1
90. Nesteruk I, Nikitin A, Shepetyuk B (2020) Study of the COVID-19 epidemic dynamics in Ukraine and its regions by methods of probabilistic and statistical analysis of evolutionary models. [Preprint.] ResearchGate. https://doi.org/10.13140/rg.2.2.30998.98884, https://www.researchgate.net/publication/343651320_Study_of_the_COVID-19_epidemic_dynamics_in_Ukraine_and_its_regions_by_methods_of_probabilistic_and_statistical_analysis_of_evolutionary_models
91. South China Morning Post (2019) https://www.scmp.com/news/china/politics/article/3044050/mystery-illness-hits-chinas-wuhan-city-nearly-30-hospitalised
92. Frankfurter Allgemeine. https://m.faz.net/aktuell/sport/mehr-sport/militaerweltspiele-2019-in-wuhan-damals-schon-corona-faelle-16758894.html
93. Data as received by WHO from national authorities, as of (2020) https://www.who.int/docs/default-source/coronaviruse/situation-reports/20201012-weekly-epi-update-9.pdf
94. Nesteruk I (2020) Нові хвилі пандемії COVID-19 в Україні. New waves of COVID-19 pandemic in Ukraine. (In Ukrainian). [Preprint.] ResearchGate. https://doi.org/10.13140/rg.2.2.26172.26244. https://www.researchgate.net/publication/343979510_Novi_hvili_pandemii_COVID-19_v_Ukraini_New_waves_of_COVID-19_pandemic_in_Ukraine
95. https://index.minfin.com.ua/ua/reference/coronavirus/ukraine/
96. https://www.kmu.gov.ua/
97. Chatterjee K et al (2020) Healthcare impact of COVID-19 epidemic in India: a stochastic mathematical model. Med J Armed Forces India 76(2):147–155
98. Ciufolini I, Paolozzi A (2020) Mathematical prediction of the time evolution of the COVID-19 pandemic in Italy by a Gauss error function and Monte Carlo simulations. Eur Phys J Plus 135:355. https://doi.org/10.1140/epjp/s13360-020-00383-y

99. Annas S et al (2020) Stability analysis and numerical simulation of SEIR model for pandemic COVID-19 spread in Indonesia. Chaos Solitons Fractals 139:110072. https://doi.org/10.1016/j.chaos.2020.110072
100. Yadav RP, Verma R (2020) A numerical simulation of fractional order mathematical modeling of COVID-19 disease in case of Wuhan China. Chaos, Solitons Fractals 140:110124. https://doi.org/10.1016/j.chaos.2020.110124
101. KY Ng, MM Gui (2020) COVID-19: development of a robust mathematical model and simulation package with consideration for ageing population and time delay for control action and resusceptibility. Physica D Nonlinear Phenomena 411:132599. https://doi.org/10.1016/j.physd.2020.132599Get
102. Ivorra B, Ferrández MR, Vela-Pérez M, Ramos AM (2020) Mathematical modeling of the spread of the coronavirus disease 2019 (COVID-19) taking into account the undetected infections. The case of China. Commun Nonlinear Sci Numer Simul 88:105303. https://doi.org/10.1016/j.cnsns.2020.105303
103. HuyTuan N, Mohammadi H, Rezapour S (2020) A mathematical model for COVID-19 transmission by using the Caputo fractional derivative. Chaos Solitons Fractals 110107. https://doi.org/10.1016/j.chaos.2020.110107
104. Sinkala M, Nkhoma P, Zulu M, Kafita D, Tembo R, Daka V. The COVID-19 pandemic in Africa: predictions using the SIR Model. medRxiv 20118893. https://doi.org/10.1101/2020.06.01.2011889
105. Agbokou1 K, Gneyou1 K, Tcharie K (2020) Investigation on the temporal evolution of the covid'19pandemic: prediction for Togo. Open J Math Sci 4:273–279. https://doi.org/10.30538/oms2020.0118. https://pisrt.org/psr-press/journals/oms
106. Pintér Gergő, Felde Imre, Mosavi Amir, Gloaguen Richard (2020) COVID-19 pandemic prediction for Hungary; a hybrid machine learning approach. Mathematics 8:890. https://doi.org/10.3390/math8060890www.mdpi.com/journal/mathematics
107. Cody Carroll et al (2020) Time dynamics of COVID-19. medRxiv 20109405. https://doi.org/10.1101/2020.05.21.2010940
108. Humphrey L et al (2020) Testing, tracing and social distancing: assessing options for the control of COVID_19. medRxiv 20077503. https://doi.org/10.1101/2020.04.23.20077503
109. Hoque E et al (2020) Adjusted dynamics of COVID-19 pandemic due to herd immunity in Bangladesh. medRxiv 20186957. https://doi.org/10.1101/2020.09.03.20186957
110. Furati KM, Sarumi IO, Khaliq AQM (2020) Memory-dependent model for the dynamics of COVID-19 pandemic. medRxiv 20141242. https://doi.org/10.1101/2020.06.26.20141242
111. Bosch J, Wilson A, O'Neil K, Zimmerman PA (2020) COVID-19Predict—predicting pandemic trends. medRxiv 20191593. https://doi.org/10.1101/2020.09.09.20191593
112. Asad A, Srivastava S, Verma MK (2020) Evolution of COVID-19 pandemic in India. medRxiv 20143925. https://doi.org/10.1101/2020.07.01.20143925
113. Breen R, Ermisch J (2020) The geography of excess deaths in England during the Covid-19 pandemic: longer term impacts and monthly dynamics. medRxiv 20188003. https://doi.org/10.1101/2020.09.07.20188003
114. Aries N, Ounis H (2020) Mathematical modeling of COVID-19 pandemic in the African continent. medRxiv 20210427. https://doi.org/10.1101/2020.10.10.20210427
115. Guenther F, Bender A, Katz K, Kuechenhoff H, Hoehle M (2020) Nowcasting the COVID-19 pandemic in Bavaria. medRxiv 20140210. https://doi.org/10.1101/2020.06.26.20140210
116. Althouse BM et al (2020) The unintended consequences of inconsistent pandemic control policies. medRxiv 20179473. https://doi.org/10.1101/2020.08.21.20179473
117. Yang W, Shaff J, Shaman J (2020) COVID-19 transmission dynamics and effectiveness of public health interventions in New York City during the 2020 spring pandemic wave. medRxiv 20190710. https://doi.org/10.1101/2020.09.08.20190710

118. Dickman R (2020) A SEIR-like model with a time-dependent contagion factor describes the dynamics of the Covid-19 pandemic. medRxiv 20169557. https://doi.org/10.1101/2020.08.06.20169557
119. Kundu LR, Ferdous MZ, Islam US, Sultana M (2020) Forecasting the spread of COVID-19 pandemic in Bangladesh using ARIMA model. medRxiv 20217414. https://doi.org/10.1101/2020.10.22.20217414
120. Carvalho D, Barbastefano R, Pastore D, Lippi MC (2020) A novel predictive mathematical model for COVID-19 pandemic with quarantine, contagion dynamics, and environmentally mediated transmission. medRxiv 20163063. https://doi.org/10.1101/2020.07.27.20163063
121. Biswas HA, Khatun MS, Paul AK, Khatun MR, Islam MA, Samad SA, Ghosh U (2020) Modeling the effective control strategy for transmission dynamics of global pandemic COVID-19. medRxiv 20076158. https://doi.org/10.1101/2020.04.22.20076158
122. Aviv-Sharon E, Aharoni A (2020) Forecasting COVID-19 pandemic severity in Asia. medRxiv 20102640. https://doi.org/10.1101/2020.05.15.20102640
123. Bannur N, Maheshwari H, Jain S, Shetty S, Merugu S, Raval A (2020) Adaptive COVID-19 forecasting via bayesian optimization. medRxiv 20215293. https://doi.org/10.1101/2020.10.19.2021529
124. Honfo SH, Taboe BH, Kakai RG (2020) Modeling COVID-19 dynamics in the sixteen West African countries. medRxiv 20188532. https://doi.org/10.1101/2020.09.04.20188532
125. Chruściel PT, Szybka SJ (2020) Universal properties of the dynamics of the Covid-19 pandemics. medRxiv 20181214. https://doi.org/10.1101/2020.08.24.20181214
126. Reddy BRM, Singh A, Srivastava P (2020) COVID-19 transmission dynamics in India with extended SEIR Model. medRxiv 20175703. https://doi.org/10.1101/2020.08.15.20175703
127. Huang J, Liu X, Zhang L, Yang K, Chen Y, Huang Z, Liu C, Lian X, Wang D (2020) The amplified second outbreaks of global COVID-19 pandemic. medRxiv 20154161. https://doi.org/10.1101/2020.07.15.2015416
128. Bhanot G, DeLisi C (2020) Analysis of Covid-19 data for eight European Countries and the United Kingdom using a simplified SIR model. medRxiv 20114058. https://doi.org/10.1101/2020.05.26.20114058
129. Ibrahim Mahmoud A, Al-Najafi Amenah (2020) Modeling, control, and prediction of the spread of COVID-19 using compartmental, logistic, and gauss models: a case study in Iraq and Egypt. Processes 8:1400. https://doi.org/10.3390/pr8111400
130. Perone G (2020) Comparison of ARIMA, ETS, NNAR and hybrid models to forecast the second wave of COVID-19 hospitalizations in Italy (HEDG-WP 20/18, University of York). Preprint. http://www.york.ac.uk/economics/postgrad/herc/hedg/wps/
131. Bueno ÁM, Batistela CM, Correa DPF, Piqueira JRC (2020) SIRSi compartmental model for COVID-19 pandemic with immunity loss. Chaos Solitons Fractals. https://doi.org/10.1016/j.chaos.2020.110388
132. Krishnamurthy A (2020) Spatiotemporal transmission dynamics of COVID-19 in Spain. Conference: advancing knowledge about spatial modeling, infectious diseases, environment and health. Presentation. https://www.researchgate.net/publication/342247591_Spatiotemporal_Transmission_Dynamics_of_COVID-19_in_Spain
133. Krishnamurthy A (2020) COVID-19 Spread in Ontario, Canada. Presentation. https://www.researchgate.net/publication/340285519_COVID-19_Spread_in_Ontario_Canada_Mar_30_to_Nov_30_2020
134. Duque D, Morton DP, Singh B, Du Z, Pasco R, Meyers LM (2020) COVID-19 in Austin, Texas: averting healthcare surges while relaxing social distancing. Preprint. https://sites.cns.utexas.edu/sites/default/files/cid/files/austin_covid-19_optimized_relaxation_of_social_distancing_0428020.pdf?m=1588082043
135. https://edition.cnn.com/2020/11/02/europe/slovakia-mass-coronavirus-test-intl/index.html

136. Coronavirus: Asian nations face second wave of imported cases (2020). Available from: https://www.bbc.com/news/world-asia-51955931. Accessed 28 July 2020
137. As China's virus cases reach zero, experts warn of second wave (2020). Available from: https://www.bloomberg.com/news/articles/2020-03-18/as-china-virus-cases-near-zero-experts-warn-of-second-wave. Accessed 28 July 2020
138. https://www.voanews.com/covid-19-pandemic/slovakias-second-round-coronavirus-tests-draws-large-crowds

The manufacturer's authorised representative in the EU is Springer Nature Customer Service Centre GmbH, Europaplatz 3, 69115 Heidelberg, Germany. If you have any concerns regarding our products, please contact ProductSafety@springernature.com

Printed and bound by CPI Group (UK) Ltd, Croydon, CR0 4YY
30/09/2026
02225306-0016